Springer Theses

Recognizing Outstanding Ph.D. Research

Aims and Scope

The series "Springer Theses" brings together a selection of the very best Ph.D. theses from around the world and across the physical sciences. Nominated and endorsed by two recognized specialists, each published volume has been selected for its scientific excellence and the high impact of its contents for the pertinent field of research. For greater accessibility to non-specialists, the published versions include an extended introduction, as well as a foreword by the student's supervisor explaining the special relevance of the work for the field. As a whole, the series will provide a valuable resource both for newcomers to the research fields described, and for other scientists seeking detailed background information on special questions. Finally, it provides an accredited documentation of the valuable contributions made by today's younger generation of scientists.

Theses are accepted into the series by invited nomination only and must fulfill all of the following criteria

- They must be written in good English.
- The topic should fall within the confines of Chemistry, Physics, Earth Sciences, Engineering and related interdisciplinary fields such as Materials, Nanoscience, Chemical Engineering, Complex Systems and Biophysics.
- The work reported in the thesis must represent a significant scientific advance.
- If the thesis includes previously published material, permission to reproduce this must be gained from the respective copyright holder.
- They must have been examined and passed during the 12 months prior to nomination.
- Each thesis should include a foreword by the supervisor outlining the significance of its content.
- The theses should have a clearly defined structure including an introduction accessible to scientists not expert in that particular field.

More information about this series at http://www.springer.com/series/8790

Takuro Sato

Transport and NMR Studies of Charge Glass in Organic Conductors with Quasi-triangular Lattices

Doctoral Thesis accepted by
the University of Tokyo, Japan

 Springer

Author
Dr. Takuro Sato
RIKEN Center for Emergent Matter Science
Dynamic Emergent Phenomena Research
  Unit
Wako, Saitama
Japan

Supervisor
Prof. Kazushi Kanoda
Department of Applied Physics
The University of Tokyo
Bunkyō, Tokyo
Japan

ISSN 2190-5053
Springer Theses
ISBN 978-981-10-5878-3
DOI 10.1007/978-981-10-5879-0

ISSN 2190-5061 (electronic)

ISBN 978-981-10-5879-0 (eBook)

Library of Congress Control Number: 2017949126

Printed on acid-free paper

This Springer imprint is published by Springer Nature
The registered company is Springer Nature Singapore Pte Ltd.
The registered company address is: 152 Beach Road, #21-01/04 Gateway East, Singapore 189721, Singapore

Supervisor's Foreword

The Ph.D. thesis of Takuro Sato is a report on his work which experimentally demonstrates that an electronic glass emerges on geometrically frustrated lattices. Glass is a ubiquitous state that appears in atomic or molecular assemblies. Such a state has been substantiated in electronic systems, as described in this thesis.

The Coulomb repulsive force between electrons makes them localized to form an electronic crystal; however, its periodicity does not always match that of the atomic or molecular lattices in materials. This situation is called geometrical frustration. If the lattice of materials is triangular and electron (or hole) density is a half per lattice point, the geometrical frustration is so strong that the electrons may not find a stable crystal but form a nontrivial state; an electronic glass is a possible state. The materials dealt with in this thesis work, layered organic compounds with anisotropic triangular lattices, are in such a situation. The hallmarks of the glass state are nonequilibrium properties, slow dynamics, and short-ranged order; all of them are captured by the experiments of charge transport, its noise, and X-ray scattering, respectively, in this work. Furthermore, the phenomena of aging, dynamical heterogeneity, and crystal growth, which have so far been exclusively the issues of soft matter physics, are observed in strongly correlated electron systems. The present work creates a novel interdisciplinary platform that integrates the science of correlated electrons and the science of soft matter. It is a great pleasure to me that Sato's thesis is published in the Springer theses series.

Tokyo, Japan
July 2017

Prof. Kazushi Kanoda

Parts of this thesis have been published in the following journal articles:

[1] F. Kagawa, T. Sato, K. Miyagawa, K. Kanoda, Y. Tokura, K. Kobayashi, R. Kumai, Y. Murakami "Charge-cluster glass in an organic conductor" *Nat. Phys.* **9**, 2642 (2013)

[2] T. Sato, F. Kagawa, K. Kobayashi, K. Miyagawa, K. Kanoda, R. Kumai, Y. Murakami, Y. Tokura "Emergence of nonequilibrium charge dynamics in a charge-cluster glass" *Phys. Rev. B* **89**, 121102(R) (2014)

[3] T. Sato, F. Kagawa, K. Kobayashi, A. Ueda, H. Mori, K. Miyagawa, K. Kanoda, R. Kumai, Y. Murakami, Y. Tokura "Systematic Variations in the Charge-Glass-Forming Ability of Geometrically Frustrated θ-(BEDT-TTF)$_2$X Organic Conductors" *J. Phys. Soc. Jpn.* **83**, 083602 (2014)

[4] T. Sato, K. Miyagawa, K. Kanoda "Fluctuation Spectroscopy Analysis Based on Dutta-Dimon-Horn Model for the Charge-Glass System θ-(BEDT-TTF)$_2$ CsZn(SCN)$_4$" *J. Phys. Soc. Jpn.* **85**, 123702 (2016)

[5] T. Sato, K. Miyagawa, K. Kanoda "Electronic crystal growth" Submitted to *Science* (under Review)

Acknowledgements

First of all, I would like to thank K. Kanoda for giving me an opportunity to work with him on this challenging and exciting project. I am grateful to him for his numerous and invaluable advice, not only on research, but also on many other aspects, such as communication and writing.

I would like to express my gratitude to K. Miyagawa for helpful discussions, continual experimental supports.

I must thank F. Kagawa for not only collaboration but also his continual encouragement, enlightening suggestions, and critical reading of my papers.

I appreciate fruitful comments and discussions from H. Tanaka, A. Maeda, Y. Iwasa, and T. Kato.

I am grateful to the excellent technical support by KEK staff, including Y. Murakami, K. Kumai, and K. Kobayashi.

I would like to thank H. Mori and A. Ueda. They provided high-quality samples for this experiment.

I also thank all the members of Kanoda group for helpful discussions and sharing fun.

Finally, I would like to thank my parents for kind encouragements. Without the unconditional support from them, this dissertation would have been impossible.

Acknowledgements

Contents

Chapter 1
Background Information

Abstract This thesis presents the experimental studies on a debated glassy behavior in charge degrees of freedom by using quasi-two-dimensional organic conductors θ-(BEDT-TTF)$_2$X. So far, this family had been considered as a typical platform of strongly correlated electron system, which exhibits metal-charge ordered insulating transition triggered by long-ranged Coulomb repulsion. In this system, however, geometrical frustration originating from triangular lattice potentially works against the long-ranged charge-ordered formation, giving rise to the possible emergence of unconventional electronic states without long-range order, namely, charge-glass states. In this chapter, we review the fundamental properties of the target systems, θ-(BEDT-TTF)$_2$X, and then noted the intriguing possibility that the family of materials shares a common physics with soft materials. We also briefly describe a general physics of glass in a classical system, focusing on a connection with the θ-(BEDT-TTF)$_2$X.

Keywords Strongly correlated electrons · Organic conductor · θ-(BEDT-TTF)$_2$X · Charge order · Charge glass · Geometrical frustration

1.1 Introduction

Interacting atoms or molecules generally condense into liquid and in turn form a crystal when cooled further; this process is a widely observed phenomenon in nature [1]. The transition into a crystal from a liquid is a first-order transition and require a finite time to be completed. In other words, crystallization is ubiquitously bypassed by cooling the system more rapidly than the characteristic time required for ordering. Such kinetic avoidance naturally leads to a non-equilibrium state; a well-established example is supercooled liquids, which generally form a glass state associated with atomic/molecular configurational degrees of freedom at lower temperatures [2, 3]. Thus, considering the framework expanded to the non-equilibrium state, the cooling speed is potentially a key parameter for determining the resulting state at low temperature. Of course, inducing disorder can also

© Springer Nature Singapore Pte. Ltd. 2017

T. Sato, *Transport and NMR Studies of Charge Glass in Organic Conductors with Quasi-triangular Lattices*, Springer Theses, DOI 10.1007/978-981-10-5879-0_1

suppress crystallization and may lead to vitrification; however, this effect is not always necessary. Indeed, a pure single-component metal liquid, which precludes any effect of extrinsic impurities, was successfully transformed into a glass state via extremely rapid quenching rates [4, 5].

However, can the idea that rapid cooling speeds generate a new state (usually hidden behind the thermodynamic ground states) be applied to the hard matter (Fig. 1.1)? When considering electrons, we have already established a corresponding "liquid", namely, the Fermi gas or Fermi liquid, in which electrons can move almost freely. Furthermore, electrons can also form a "crystal" via strong Coulomb interactions, which is often called a Wigner crystal or charge ordering (CO). In CO states, an equal number of charge-rich and charge-poor sites alternatively occupy a lattice, breaking the translational symmetry [6]. A glass state in electrons that is between a "charge liquid" and a "charge crystal", however, has not been found to date. Note that the several inhomogeneous states that are often highlighted in some electronic systems close to the metal-insulator transition [7–9] cannot be the pure electronic glass states we seek. In these systems disorder instead of electronic correlation is revealed to be associated strongly with the resultant inhomogeneity. The emergence of the charge-glass state, even in the absence of disorder, is thus one of the major scientific challenges, and is of fundamental importance for better understanding both strongly correlated electron system and the general physics of glassiness.

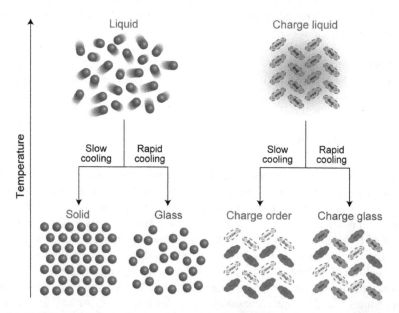

Fig. 1.1 Schematics of the cooling-rate-dependent solidification and vitrification of liquids. The *left* figures represent the classical system, whereas the *right* figures represent the quantum system

In this thesis, this issue was addressed using organic conductors, θ-(BEDT-TTF)$_2$X family of materials, which are typically used for the study of charge-ordered physics. We aimed to establish experimentally the concept of the novel glass in charge degrees of freedom.

1.2 Organic Conductors

We begin by describing the basic characteristics of organic conductors, which exhibit a variety of interesting electronic properties. Generally most of organic conductors have the following key advantages that are of interest; (i) simple band structure, (ii) low dimensionality, and (iii) strong electron correlation.

(i) Simple band structure

Although organic conductors consist of a large number of atoms, the resulting electronic band structure is relatively simple. This simple structure occurs because of the relatively longer distance between the molecules via van der Waals (vdW) interaction, resulting in smaller *inter*-molecular transfer integrals than the *intra*-molecular ones. Thus, the *intra*-molecular transfer integral forms a set of non-degenerate molecular orbitals (MOs); each MO then connects together via small *inter*-molecular ones. In such cases, only the HOMO (highest occupied molecular orbital) plays the decisive role in the electronic properties [10]. Indeed, the tight-binding calculation based on the *inter*-molecular overlap of the HOMO band agrees very well with ab initio calculations.

(ii) Low dimensionality

The building units of the HOMO are the π orbitals of the molecules with almost flat structure. Thus, organic conductors generally tend to have a highly anisotropic character, generating a low dimensionality in the electron nature based on the molecular type; for example, DCNQI or TMTTF compounds show a one-dimensional nature because of their chain structures, whereas BEDT-TTF compounds, which are the target materials of our study, show a two-dimensional nature due to their layered structures.

(iii) Strong electron correlation

The electronic bandwidth of organic conductors is smaller by approximately one order of magnitude compared to that of inorganic compounds. This difference is another consequence of the weak vdW coupling between molecules. The typical bandwidth value for organic systems is known to be on the order of the Coulomb interaction energy, resulting in relatively strong electron-electron Coulomb interactions. Indeed, typical insulating states, such as the Mott insulator or the charge-ordered insulator, caused by strong correlation are often observed in the BEDT-TTF family of materials [11, 12].

1.3 θ-(BEDT-TTF)₂X

Here, we review the fundamental properties of the target systems, θ-(BEDT-TTF)₂X. In this thesis, three materials with different anion X molecules were studied: θ-(BEDT-TTF)₂CsZn(SCN)₄, θ-(BEDT-TTF)₂RbZn(SCN)₄, and θ-(BEDT-TTF)₂TlCo(SCN)₄ (hereafter denoted as θ-CsZn, θ-RbZn, and θ-TlCo, respectively). The crystal structure and previous experimental results are shown below.

1.3.1 Crystal Structure

θ-(BEDT-TTF)₂X are a charge-transfer type organic compounds consisting of cationic BEDT-TTF molecules and anionic X molecules, where BEDT-TTF (abbreviated as ET) denotes bis(ethylenedithio)tetrathiafulvalene. Their crystal structure is composed of alternating layers, conducting ET layers and insulating anion X layers, resulting in a quasi-two-dimensional electron nature (Fig. 1.2a) [13, 14]. Anion X is fully charged as a −1 with closed shell formed; thus, the average charge of ET molecules is +1/2. In other words, one hole per two ET molecules is accommodated by the anion layers. Consequently, the resulting conduction band originating from ET molecules is 3/4-filled (or 1/4-filled in terms of holes) if all the ET molecules are equivalent. The Greek character "θ" represents the specific packing pattern of the ET molecules, as shown in Fig. 1.2b. The ET molecules form anisotropic triangular lattices in the conducting layers. According to the symmetry, there are two ET molecules in the unit cell, which are crystallographically equivalent (space group *I222*), and two types of transfer integrals, t_c and t_p (Fig. 1.3a) [13, 14].

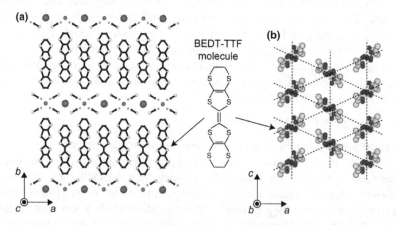

Fig. 1.2 Crystal structure of θ-(BEDT-TTF)₂X system. **a** Layered structure of θ-(BEDT-TTF)₂X. The two-dimensional conducting layers of BEDT-TTF molecules are separated by insulating anion layers. **b** The structure of conducting layer, in which BEDT-TTF molecules form anisotropic triangular lattices. The inset presents the structure of the BEDT-TTF molecule

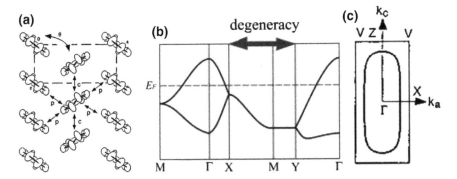

Fig. 1.3 a Map of the transfer integrals in the conducting plane. **b** The tight-binding band structure. **c** The Fermi surface obtained from (**b**)

In Figs. 1.3b, c, the tight-binding band structure and the Fermi surface of θ-type are shown. There are two bands in the Brillouin zone because of the two molecules in the unit cell, and the Fermi level is located in the higher band. Clearly, the two bands along the Brillouin zone boundary are degenerate; hence, the bands as a whole are considered to be 3/4-filled.

In this band filling condition, long-ranged Coulomb repulsion among electrons generally forces them to form an insulating CO state. Mean-field calculations based on the extended Hubbard model, which includes the term of inter-site Coulomb repulsion, V, as well as that of on-site Coulomb repulsion, U, also support the emergence of the CO state. In triangular lattices, however, "geometrical frustration" potentially works against the formation of the long-ranged CO state by prohibiting the unique configuration of charge as discussed in the next section. Such a competing effect is the key to understanding the fascinating electronic properties of the θ-(BEDT-TTF)₂X family.

1.3.2   Geometrical Frustration

Frustration refers to the existence of competing forces that cannot be simultaneously satisfied. We first introduce the concept of geometrical frustration by using a well-known Ising-type spin system of triangular lattices, where the nearest-neighbor interactions favor anti-aligned spins [15]. As clearly shown in Fig. 1.4a, the three spins cannot all be antiparallel on a triangle, giving rise to a very large degeneracy of ground states. Currently, the physics of a strongly correlated electron system attaches much importance to degeneracy, which leads to the possible emergence of exotic electronic states without long-ranged ordering, such as quantum spin liquids [15], or classically disordered states, such as spin ices [16, 17] or spin glasses [18]. Note that the BEDT-TTF systems with 1/2-filled bands host an antiferromagnetic

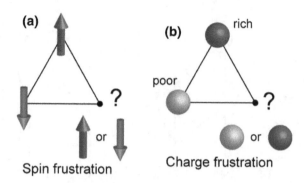

Fig. 1.4 An illustration of the analogy of **a** spin frustration and **b** charge frustration originating from a triangular lattice

condition consisting of a two-dimensional triangular lattice, serving as one of the prototypes of the frustrated spin system.

Thus, we apply the analogous approach to the charge-ordered systems with a 1/4-filled band. Considering that the charge-rich and charge-poor sites correspond to the up and down spins, respectively, we can expand the concept of geometrical frustration on the triangular lattices into the charge-ordered systems (Fig. 1.4b). Wigner-type charge ordering is a phenomenon in which an equal number of charge-rich and charge-poor sites occupy a lattice such that rich–rich (or poor–poor) neighboring pairs are avoided as much as possible. However, in the triangular lattice, this constraint is insufficient to determine a specific charge ordering among the various charge configurations. Thus, the "charge" frustration potentially undermines the tendency toward long-ranged charge ordering, as was first suggested in Ref. [19]. Considering the intriguing examples in the spin systems, unconventional electronic states may also be exhibited at low temperatures when long-range charge ordering is avoided [20].

The θ-(BEDT-TTF)$_2$X system turns out to host the charge frustrations corresponding to triangular lattices in the conducting layers. Based on the structure of θ-phase materials, T. Mori calculated the energy of possible charge-ordered patterns within the static limit, that is, the ideal conditions when the transfer integrals in it are neglected [21, 22]. He proposed that the CO patterns shown in Fig. 1.5 are

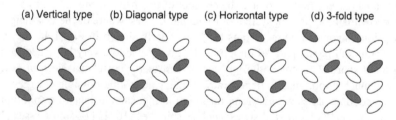

Fig. 1.5 Possible charge-ordered patterns of the θ-phase system. **a** *Vertical*, **b** *Diagonal*, and **c** *Horizontal* type stripes, **d** 3-fold pattern. The 3-fold type of charge configuration is not of stripe order, and is found to be degenerate with the three stripe patterns

almost degenerate in energy, and compete with each other. Note that, in addition to the typical stripe-type CO patterns (Figs. 1.5a, b, and c), the non-stripe CO configuration called 3-fold type (Fig. 1.5d), is also degenerate, in line with the experimental results for θ-(BEDT-TTF)₂X systems as explained below.

1.3.3   *Previous Experimental Studies of θ-(BEDT-TTF)₂X*

In this subchapter, we briefly summarize the results of previous experimental studies of θ-RbZn, θ-CsZn, and θ-TlCo. Although all these materials have similar structures and the same symmetry, their resulting transport properties are markedly different from each other as shown in Fig. 1.6. In the last part of this section, we propose a new approach for gaining a systematic understanding of the series of results, which is based on the notion of geometrical frustration.

θ-(BEDT-TTF)₂RbZn(SCN)₄

We begin by describing the properties of θ-RbZn, which exhibits the most remarkable temperature-resistivity profile among the materials we used. At room temperature, the charges are delocalized in the conducting layer; thus, the charge density is considered to be homogeneous. Thus, we refer to this electronic state as a "charge liquid". When slowly cooled (<1 K/min), the system shows a sudden increase in resistivity at approximately 200 K with a clear hysteresis, concluding the first-ordered phase transition. Several experiments confirmed that the transition is due to the formation of the horizontal type of the charge-ordered (or "charge crystal") state, with substantial structural modulation of $q = (0, 0, 1/2)$, in

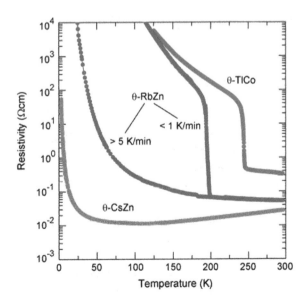

Fig. 1.6 Temperature dependence of the resistivity for three different materials: θ-TlCo (*red*), θ-RbZn (*blue*), and θ-CsZn (*green*)

accordance with the general nature of the 1/4-filled band structure [23–25]. The structural modulation releases the degree of charge frustration, stabilizing the long-range CO states. The resultant crystal structure below T_{CO} is in the form of orthorhombic $P2_12_12_1$, and we call it θ_d-type to distinguish it from that of the high-temperature charge-liquid state. Interestingly, when rapidly cooled (>4 K/min), the charge/structural ordering at 200 K vanishes, providing an alternative approach to achieve a high conducting state, with frustrated triangular lattices maintained even in the temperature regime below 200 K [26, 27]. Because there is no resistivity anomaly at 200 K during rapid cooling, the rapidly cooled phase below 200 K is considered a continuation of the electronic state above 200 K, strongly reminiscent of supercooled state widely observed in viscous liquids. Furthermore, the electronic states above 200 K may be exotic, as revealed through several experiments; for example, the optical conductivity measurements revealed that the high-temperature charge-liquid states cannot be ascribed to conventional metallic states that have a well-defined Drude response [28]. More interestingly, previous NMR measurements indicated the existence of anomalous slow charge dynamics of the order of kilohertz above 200 K [29]. Because these phenomena are key to understanding the overall properties of θ-(BEDT-TTF)$_2$X, we next show the results of NMR measurements in detail.

Figure 1.7a displays the temperature dependence of ^{13}C-NMR spectra at the so-called Magic angle (see Chap. 2) for slowly cooled θ-RbZn [29]. At room temperature, the spectrum consists of two narrow lines, reasonably explained by the

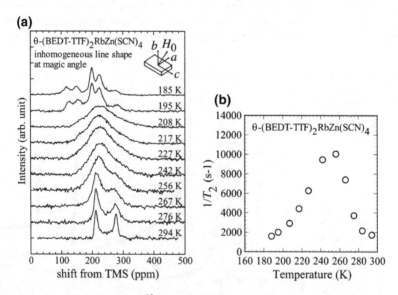

Fig. 1.7 The previous results of ^{13}C-NMR for slowly cooled θ-RbZn [29]. **a** Temperature dependence of the NMR spectra. The applied field is precisely tuned at the so-called magic angle. **b** Temperature dependence of the spin-spin relaxation rate, $1/T_2$, which proves the slow fluctuations are of the order of kilohertz

two adjacent ^{13}C atoms in the center of the BEDT-TTF molecule. This result indicates that the electronic state at room temperature is homogenous. As the temperature is lowered, however, the spectrum becomes broadened, indicating a continuous distribution in charge density with slow charge fluctuations distinctive from conventional metallic states [29]. The formation of long-range charge ordering below T_{CO} is also confirmed clearly; when the temperature slowly crosses 200 K, the spectrum changes its shape into two components, two sharp peaks and broad one, originating from the charge-poor and charge-rich sites, respectively.

In Fig. 1.7b, the spin-spin relaxation rate, $1/T_2$, is plotted against temperature, which is a sensitive indicator of the slow dynamics of the order of kilohertz. The enhancement of $1/T_2$ at approximately 250 K suggests that the dynamics in charge slows down and passes the window of kilohertz order at approximately 250 K. The temperature dependence of magnetic susceptibility for θ-RbZn (shown in Fig. 1.8) also supports the existence of slow charge dynamics [13]. In the wide temperature range from room temperature to 50 K, magnetic susceptibility shows little temperature dependence, suggesting that spins behave almost paramagnetically. In addition, there exists no anomaly at approximately 250 K, eliminating the possibility that spin fluctuations contribute to the enhancement of $1/T_2$. All these results indicate that θ-RbZn exhibits anomalous slow dynamics in the charge sector with spatial inhomogeneity in the high-temperature metallic region. Therefore, the electronic state when it is cooled rapidly is possibly considered as a "supercooled" or even "glassy" state in terms of charge.

Finally, we refer to the previous X-ray scattering measurements, which also revealed the non-trivial behavior in the metallic region [24]. Above T_{CO}, diffuse planes and diffuse rods were found around the Bragg reflections as shown in

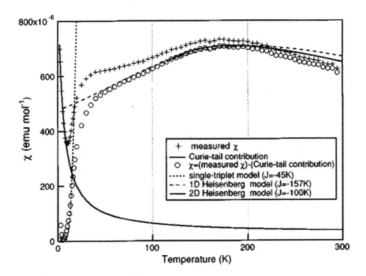

Fig. 1.8 Temperature dependence of the spin susceptibility for slowly cooled θ-RbZn [11]

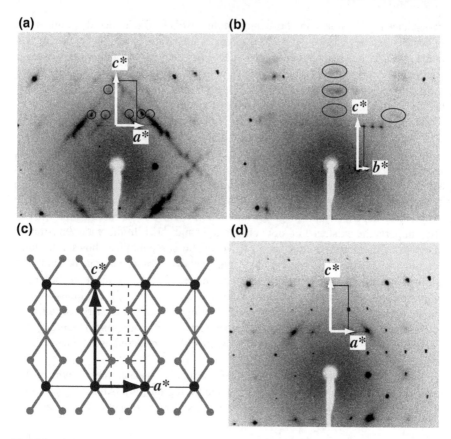

Fig. 1.9 Oscillation photographs around T_{CO} [21]. **a** and **b** were taken at 220 K above T_{CO}. **c** Schematics of the diffuse planes and the diffuse rods around the Bragg reflections above T_{CO}. **d** Superlattice reflections of horizontal CO at 190 K

Fig. 1.9a. The diffuse rods are characterized by the wave vector of $q_1 = (1/4, k, 1/3)$, being different from the that of horizontal CO, $q_2 = (0, 0, 1/2)$. This result indicates that short-ranged CO domains with q_1 not with q_2 may be coupled with the slow fluctuations detected by NMR. The q_1 type of short-ranged CO domains suddenly vanish when the CO transition is realized below T_{CO}, and instead the superlattice reflections with q_2 clearly appear, confirming the formation of long-ranged horizontal type CO (Fig. 1.9d).

θ-(BEDT-TTF)$_2$CsZn(SCN)$_4$

In contrast to θ-RbZn, θ-CsZn shows no signature of first-ordered CO transition in resistivity even when it is cooled slowly, and the resistivity starts to increase gradually below approximately 50 K (Fig. 1.6). Infrared and Raman studies showed that the average space group symmetry remains $I222$ from room temperature down to 6 K, consistent with the absence of long-ranged CO formation [28].

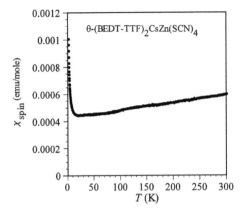

Fig. 1.10 Temperature dependence of the spin susceptibility for θ-CsZn [30]

The spin susceptibility, χ_{spin}, against temperature shown in Fig. 1.10 is almost temperature independent down to 20 K, which is reminiscent of Pauli paramagnetism [13, 30]. ^{13}C-NMR measurements were also performed for θ-CsZn, and the temperature dependence of spectra is shown in Fig. 1.11a [31]. There is no drastic

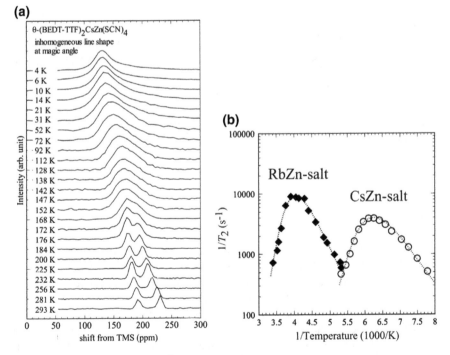

Fig. 1.11 The previous results of ^{13}C-NMR for θ-CsZn [31]. **a** Temperature dependence of the NMR spectra. The applied field is precisely tuned at the so-called magic angle. **b** Temperature dependence of the spin-spin relaxation rate, $1/T_2$. Also shown is $1/T_2$ for θ-RbZn

Fig. 1.12 Oscillation
photograph for θ-CsZn at
33.5 K [32]

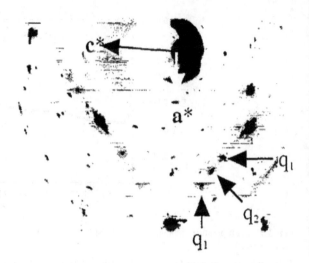

change in spectra, confirming the absence of long-ranged CO transition over the
temperature range studied. The center of mass of the spectrum exhibits a small
temperature dependence, which appears to be well scaled to that of χ_{spin}. Note that
the spectrum becomes broadened towards lower temperatures (below 160 K) and
the spin-spin relaxation rate, $1/T_2$, simultaneously increases, suggesting the
emergence of inhomogeneous charge density with slow fluctuations similar to that
of the charge-liquid state in θ-RbZn (Fig. 1.11b, in which open circles and closed
squares indicate the results of the θ-CsZn salt and the θ-RbZn, respectively). All
these results strongly indicate that electronic state of θ-CsZn is qualitatively similar
to that of rapidly cooled θ-RbZn.

Unlike θ-RbZn, however, θ-CsZn possesses two different short-ranged CO
domains characterized by $q_1 = (2/3, k, 1/3)$ and $q_2 = (0, k, 1/2)$ simultaneously,
indicating that they compete with each other (Fig. 1.12) [32]. Below 50 K, a distinct increase only in the intensity of q_2 wave vector is found, possibly contributing
to the increase in resistivity at the same temperature range.

1.3.3.1 θ-(BEDT-TTF)$_2$TlCo(SCN)$_4$

Finally, we describe the electronic properties for θ-TlCo. In the temperature-
resistivity profile for θ-TlCo, a distinct resistivity jump was observed at 245 K. This
transition is confirmed to be due to horizontal CO formation by infrared and Raman
spectroscopy [13, 14, 33]. At the same time, the crystal structure is modulated from
the space group of $I222$ to that of $P2_12_12_1$, being the same as that of slowly cooled
θ-RbZn. It is noted, however, that a pronounced cooling dependent behavior
observed in θ-RbZn has not been experimentally found, to date.

1.3.4   Previous Phase Diagram for θ-(BEDT-TTF)$_2$X

H. Mori and co-workers proposed the phase diagram for the θ-(BEDT-TTF)$_2$X system in terms of the dihedral angle, defined as the angle between the molecules in adjacent columns [13]. They indicated that the transfer integral in the transverse direction decreases as the dihedral angle increases according to the extended Huckel calculation. As the transfer integral decreases, the effect of the Coulomb interaction will be relatively dominant, thus stabilizing the CO states. However, an increase in the dihedral angle also leads to an increase in the transfer integral in the vertical direction, which possibly works against the formation of resulting horizontal CO states. Furthermore, the distinct cooling-rate dependent electronic state or exotic slow dynamics in charge observed experimentally, which may be crucially significant for the understanding of overall nature of θ-(BEDT-TTF)$_2$X, is not describable in Fig. 1.13. These counterexamples suggest the existence of another important parameter.

1.3.5   New Possibility in Terms of Geometrical Frustration

As described above, the θ-(BEDT-TTF)$_2$X system shows a variety of physical properties, but appears to vary systematically within the series of the isostructural compounds; the CO state seemingly becomes less stable and instead a possible supercooled electronic state is realized in the order of θ-TlCo, θ-RbZn, and θ-CsZn. This tendency provides insights into which material parameter plays a key role in the resultant electronic states. We here aim to explain the series of data in terms of the underlying geometrical charge frustration in θ-(BEDT-TTF)$_2$X. To estimate the degree of the charge frustration, we adopt the simplest parameter, V_p/V_c, i.e., the ratio of the two different Coulomb interactions. In this family, the strength of charge frustrations varies systematically depending on anion X; for example, the values of V_p/V_c for θ-CsZn, θ-RbZn, and θ-TlCo are approximately 0.92, 0.87, and 0.84, respectively [21]. Therefore, we hypothesize that stronger charge frustration tends to make a CO state more unstable, leading to the formation of a supercooled or CG state. Such a viewpoint is thought to be natural in the physics of soft materials, especially in the glass-forming system.

1.4   Classical Glass

In the previous section, we noted the intriguing possibility that the θ-(BEDT-TTF)$_2$X family shares a common physics with soft materials, especially in the nature of the glass state. Here, we briefly review "glassiness" in the classical system, focusing on the connection with θ-(BEDT-TTF)$_2$X. In this section, the

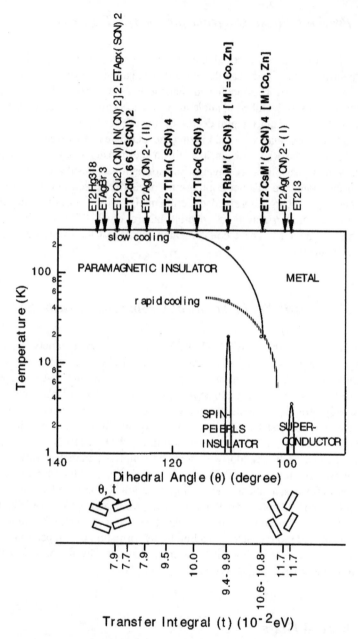

Fig. 1.13 Phase diagram for θ-type BEDT-TTF salts as a function of the transfer integral in the transverse direction and the dihedral angle of the donor columns [13]

glass state is revealed to be one of the typical nonequilibrium states, being clearly distinguishable from a gas, liquid, or solid state and realized naturally in the paradigm of supercooled liquid. Therefore, we first review the basic concept of supercooled liquids and then describe the general features of the glass state.

1.4.1 Phenomenology of Supercooled Liquids

Figure 1.14 represents the general temperature dependence of a liquid's volume (or enthalpy) [2, 34]. Upon cooling, a liquid generally becomes a crystal, that is, a thermodynamic ground state, at the melting point, T_m. Notably, this first-ordered transition can be kinetically prevented by thermal quenching [35, 36], giving an alternative pathway to the supercooled state, whereby a liquid state can persist even below T_m. In the supercooled regime, the dynamics of structural relaxation becomes extremely slow at lower temperatures, and eventually exceeds the time scale of our experimental observation at a certain temperature, T_g. In other words, molecules cannot search adequate configurations in the time allowed by the cooling rate at T_g. This is the glass transition. As a result, this transition is often described as an ergodic to non-ergodic transition. This transition is not a true thermodynamic phase transition, as it does not exhibit discontinuous changes in any physical properties. Reflecting the kinetic origin of the glass transition, the experimental T_g^* can be tunable by the cooling rate distinctive from the thermodynamic transition; faster cooling leads to a higher T_g^* (see Fig. 1.14) [37].

Fig. 1.14 General temperature dependence of volume (or enthalpy) for liquids [2, 34]

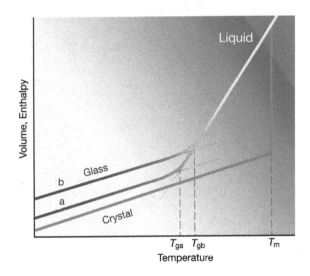

1.4.2 Structural Dynamics in Glass-Forming Systems

One of the most striking features of glass-forming systems is a dramatic slowing down in the structural relaxation time, τ, with decreasing temperature. Figure 1.15 shows the Arrhenius plot of the relaxation time (or viscosity) scaled by T_g for a range of glass-forming materials (so-called an "Angell plot") [2, 3, 36, 38]. This representation of the relaxation time provides a useful method to compare the slowing-down manner for different liquids. On the one hand, materials such as SiO_2 or GeO_2, which show an Arrhenius type of the function, $\tau = \tau_0 \exp(\Delta E/k_B T)$, in the growth of the relaxation time between T_g and the high-temperature limit, are categorized as "strong glass". The other materials such as o-terphenyl, on the other hand, exhibit a strongly non-Arrhenius temperature dependence as T_g is approached, and can be well described by the Vogel-Tammann-Fulcher (VTF) equation: $\tau = A \exp[B/(T - T_0)]$, where A and B are temperature-independent constants [39–41]; these materials are known as "fragile glass". Such a classification is crucially significant in studying the origin of glassy dynamics. In the strong glass, a particular energy scale, ΔE, dominates structural rearrangement over a whole temperature region. In other words, there exists no cooperative dynamics in the strong glass. Alternatively, the VTF form of relaxation time in the fragile glass indicates that an effective energy gap is larger at lower temperatures because of the presence of

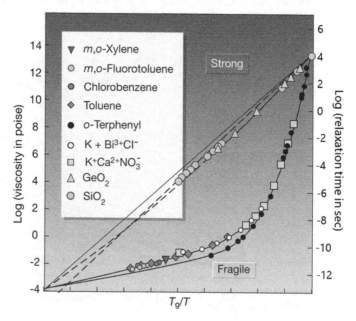

Fig. 1.15 T_g-scaled Arrhenius plot of a liquid's relaxation time or viscosities

an increasing number of dynamically correlated molecules [42]. The degree of fragility can be quantitatively evaluated by the fragility index, m, defined as

$$m = \frac{d(\log \tau)}{d(T_g/T)}\bigg|_{T=T_g},$$

which corresponds to the normalized energy gap at approximately T_g [43]. The larger value of m indicates more noticeable deviations from the Arrhenius behavior and more fragile liquids. The typical value of m for the strong liquid (e.g., SiO_2) decreases to approximately 20, whereas that for the fragile liquid (e.g., o-terphenyl) reaches approximately 100 [44].

1.4.3 Time-Dependent Relaxation Process

Next, we focus on the time dependence of the relaxation processes. The glass-forming system close to T_g often shows nonexponential decay in response to external perturbation. The response function, $F(t)$, is often described by the well-known Kohlrausch-Williams-Watts (KWW) function [45, 46]:

$$F(t) = \exp\left[-(t/\tau)^\beta\right](0<\beta\leq 1),$$

where τ is the characteristic relaxation time at the fixed temperature, and β is the stretching exponent indexing the nonexponentiality. The case of single exponential relaxation is given by $\beta = 1$, whereas an increase in nonexponentiality is represented by a decreasing value of β. The value of β usually decreases at lower temperatures (Fig. 1.16) [47]. The origin of the nonexponentiality of the dynamics is a hotly debated topic. Bohmer et al. analyzed the nonexponential relaxation in a large number of strong and fragile glass formers, suggesting a general, relatively broad correlation between fragility and nonexponentiality (Fig. 1.17) [44]. More importantly, recent results of experimental and computational studies suggest that the nonexponential dynamics are coupled to spatial heterogeneity [48–52]. This picture assumes the heterogeneous set of environments, in which each relaxation process is nearly exponential, but the relaxation time strongly depends on the given environments, resulting in an entirely nonexponential relaxation (Fig. 1.18) [53]. Within this picture, the slowing down in τ and decreasing β is caused by the growth of heterogeneous domains and the enhanced heterogeneity, respectively. Note that the question of whether such a heterogeneous scenario is relevant or not remains controversial, and an alternative explanation that the system is homogeneous but each relaxation process is intrinsically nonexponential is also discussed [54]. However, the picture of the heterogeneous domain is still valuable, providing an intuitive explanation of glassy freezing.

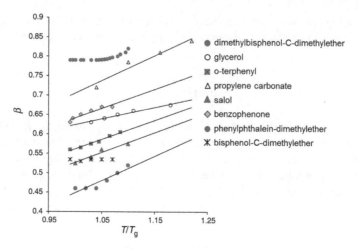

Fig. 1.16 Temperature dependence of the stretching exponent β for the several glass-forming materials, obtained from Ref. [47]

Fig. 1.17 Fragility, m, plotted against the stretching exponent, β, for the several glass formers

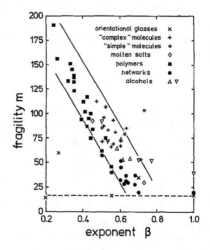

1.4.4 Critical Cooling Rate

As described above, the glass transition is ubiquitous in nature; thus, in principle all liquids can be vitrifiable by sufficiently rapid cooling. Within the accessible cooling speed in laboratories, however, only a limited set of families of materials can be successfully transformed into a glassy state, and crystallization is inevitable in the other systems. Furthermore, even for the liquids that have been successfully transformed into a glassy state, the minimum cooling speed required for the kinetic avoidance of crystallization, namely, critical cooling rate, R_c, is widely distributed by more than 10 orders of magnitude (Fig. 1.19). In the field of oxide glasses or

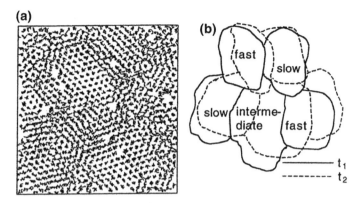

Fig. 1.18 Images of spatially heterogeneous dynamics [53]. **a** Overlaid images of particle positions at various times. Regions where distinct particles are easily visible have a fixed structure over this time period. Other regions show substantial rearrangement of local structure on the same timescale. **b** Schematic illustration of regions of spatially heterogeneous dynamics near T_g

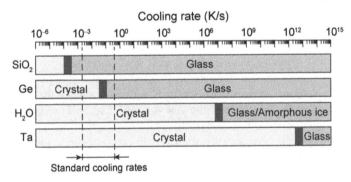

Fig. 1.19 Resulting states as a function of cooling rate for several viscous liquids [55]

metallic glasses, the material dependence of R_c is often discussed in terms of "glass-forming ability", which is evaluated quantitatively from the inverse of R_c. The search for what dominates the glass-forming ability is one of the key interests of the physics of glass.

1.5 Purpose of the Thesis

As mentioned above, the organic conductor θ-(BEDT-TTF)$_2$X family exhibits many exotic phenomena, such as cooling-rate-dependent transport properties, inhomogeneous charge density, and slow charge fluctuations. All these behaviors are strongly reminiscent of classical glass in soft materials, suggesting the existence of the glass state in charge degrees of freedom. The electronic nature of the θ-(BEDT-TTF)$_2$X family is possibly coupled with the geometrical charge frustration originating from triangular lattices. Thus, the presence of a novel glass state in electrons is not unlikely, analogous to unconventional spin states without long-range order, such as spin liquids or spin glasses driven by geometrical spin frustration. The question of whether a "charge-glass" state can be realized or not has attracted much interest to date; however, no one has experimentally demonstrated this phenomenon, and the issue remains an open question.

Therefore, the main purpose of our study was to experimentally demonstrate the existence of glass in electrons. We explored the organic conductor θ-(BEDT-TTF)$_2$X family of materials, focusing our full attention on capturing their CG behavior. We used a combination of conventional resistivity measurements, time-resolved transport measurements, and X-ray diffraction measurements to characterize the various hallmarks of glass. If the presence of a CG state is successfully demonstrated, we then perform systematic studies of the three isostructural materials with different strengths of charge frustration. This inspection can provide a clue to the correlation between charge frustration and CG behaviors, which is possibly connected to the general physics of glass. Through the set of investigations above, we aim to gain a comprehensive understanding of the electronic state under geometrical charge frustration as well as establish the universal scenario in glassiness within the strongly correlated electron system on an equal footing with soft materials. This is the ultimate goal of our study.

This thesis is organized as follows. In Chap. 2, we describe the materials and experimental methods used in our study. Chapter 3 presents the experimental demonstrations for the existence of the charge-glass state, and the last part of this chapter is devoted to the discussion of the effect of the geometrical frustration. In Chap. 4, we report the observations of electronic crystal growth. Finally, we conclude the present work in Chap. 5.

References

1. U. Gasser, J. Phys. Condens. Matter **21**, 203101 (2009)
2. P.G. Debenedetti, F.H. Stillinger, Nature **410**, 259 (2001)
3. C.A. Angell, Science **267**, 1924 (1995)
4. L. Zhong, J. Wang, H. Sheng, Z. Zhang, S.X. Mao, Nature **512**, 177 (2014)
5. J. Orava, A. L. Greer, J. Chem. Phys. **140**, (2014)
6. E. Wigner, Phys. Rev. **46**, 1002 (1934)
7. E. Dagotto, Science **309**, 257 (2005)
8. A. Vaknin, Z. Ovadyahu, M. Pollak, Phys. Rev. Lett. **84**, 3402 (2000)
9. S. Bogdanovich, D. Popović, Phys. Rev. Lett. **88**, 236401 (2002)
10. K. Fukui, F. Hiroshi, *Frontier Orbitals and Reaction Paths* (World Scientific, 1997)
11. K. Miyagawa, K. Kanoda, A. Kawamoto, Chem. Rev. **104**, 5635 (2004)
12. H. Seo, C. Hotta, H. Fukuyama, Chem. Rev. **104**, 5005 (2004)
13. H. Mori, S. Tanaka, T. Mori, Phys. Rev. B **57**, 12023 (1998)

14. H. Mori, S. Tanaka, T. Mori, A. Kobayashi, H. Kobayashi, Bull. Chem. Soc. Jpn **71**, 797 (1998)
15. L. Balents, Nature **464**, 199 (2010)
16. A.P. Ramirez, A. Hayashi, R.J. Cava, R. Siddharthan, B.S. Shastry, Nature **399**, 333 (1999)
17. S.T. Bramwell, Science **294**, 1495 (2001)
18. J.P. Bouchaud, L.F. Cugliandolo, J. Kurchan, M. Mezard, in *Spin Glasses and Random Fields* (1997), pp. 161–223
19. P.W. Anderson, Phys. Rev. **102**, 1008 (1956)
20. J. Merino, H. Seo, M. Ogata, Phys. Rev. B **71**, 125111 (2005)
21. T. Mori, J. Phys. Soc. Jpn. **72**, 1469 (2003)
22. H. Seo, J. Phys. Soc. Jpn. **69**, 805 (2000)
23. K. Miyagawa, A. Kawamoto, K. Kanoda, Phys. Rev. B **62**, R7679 (2000)
24. M. Watanabe, Y. Noda, Y. Nogami, H. Mori, J. Phys. Soc. Jpn. **73**, 116 (2004)
25. K. Yamamoto, K. Yakushi, K. Miyagawa, K. Kanoda, A. Kawamoto, Phys. Rev. B **65**, 85110 (2002)
26. F. Nad, P. Monceau, H.M. Yamamoto, Phys. Rev. B **76**, 205101 (2007)
27. Y. Nogami, N. Hanasaki, M. Watanabe, K. Yamamoto, T. Ito, N. Ikeda, H. Ohsumi, H. Toyokawa, Y. Noda, I. Terasaki, H. Mori, T. Mori, J. Phys. Soc. Jpn. **79**, 44606 (2010)
28. H. Tajima, S. Kyoden, H. Mori, S. Tanaka, Phys. Rev. B **62**, 9378 (2000)
29. R. Chiba, K. Hiraki, T. Takahashi, H.M. Yamamoto, T. Nakamura, Phys. Rev. Lett. **93**, 19 (2004)
30. T. Nakamura, W. Minagawa, R. Kinami, J. Phys. Soc. Jpn. **69**, 504 (2000)
31. R. Chiba, K. Hiraki, T. Takahashi, H.M. Yamamoto, T. Nakamura, Phys. Rev. B **77**, 1 (2008)
32. Y. Nogami, J.-P. Pouget, M. Watanabe, K. Oshima, H. Mori, S. Tanaka, T. Mori, Synth. Met. **103**, 1911 (1999)
33. K. Suzuki, K. Yamamoto, K. Yakushi, Phys. Rev. B **69**, 85114 (2004)
34. P.G. Debenedetti, *Metastable Liquids: Concepts and Principles* (Princeton University Press, 1996)
35. D. Turnbull, Contemp. Phys. **10**, 473 (1969)
36. C.A. Angell, J. Non-Cryst. Solids **102**, 205 (1988)
37. R. Brüning, K. Samwer, Phys. Rev. B **46**, 11318 (1992)
38. J.L. Green, K. Ito, K. Xu, C.A. Angell, J. Phys. Chem. B **103**, 3991 (1999)
39. H. Vogel, Phys. Z. **22**, 645 (1921)
40. G. Tammann, W. Hesse, Z. Für Anorg. Und Allg Chemie **156**, 245 (1926)
41. G.S. Fulcher, J. Am. Ceram. Soc. **8**, 339 (1925)
42. M.D. Ediger, Annu. Rev. Phys. Chem. **51**, 99 (2000)
43. C.A. Angell, K.L. Ngai, G.B. McKenna, P.F. McMillan, S.W. Martin, J. Appl. Phys. **88**, 3113 (2000)
44. R. Böhmer, K.L. Ngai, C.A. Angell, D.J. Plazek, J. Chem. Phys. **99**, 4201 (1993)
45. R. Kohlrausch, Ann. Der Phys. Und Chemie **167**, 56 (1854)
46. G. Williams, D.C. Watts, Trans. Faraday Soc. **66**, 80 (1970)
47. A.F. Kozmidis-Petrović, J. Therm. Anal. Calorim. (2016)
48. M.T. Cicerone, M.D. Ediger, J. Chem. Phys. **103**, 5684 (1995)
49. M.T. Cicerone, M.D. Ediger, J. Chem. Phys. **104**, 7210 (1996)
50. D.N. Perera, P. Harrowell, J. Chem. Phys. **104**, 2369 (1996)
51. R. Böhmer, G. Hinze, G. Diezemann, B. Geil, H. Sillescu, Europhys. Lett. **36**, 55 (1996)
52. L. Berthier, G. Biroli, J. Bouchaud, L. Cipelletti, F. Ladieu, M. Pierno, D. El Masri, D.L. Ho, Science **310**, 1797 (2005)
53. M. Hurley, P. Harrowell, Phys. Rev. E **52**, 1694 (1995)
54. M.D. Ediger, Annu. Rev. Phys. Chem. **51**, 99 (2000)
55. F. Kagawa, H. Oike, Adv. Mater. **29**, 1601979 (2016)

Chapter 2
Experimental

Abstract Charge-order or charge-glass states are in principal characterized by charge configuration on lattices, offering the opportunity to probe their dynamics by charge-sensitive measurements. It is an experimental advantage for studying glassiness in charge, which is distinctive from the case of the atomic or molecular systems. In this chapter, four different methods we performed are described; (i) resistivity measurements, (ii) resistance fluctuation spectroscopy (so-called noise measurements), (iii) X-ray diffuse scattering measurements, and (iv) NMR measurements.

Keywords Resistivity measurements · Resistance fluctuation spectroscopy · X-ray diffuse scattering measurements · NMR measurements

2.1 Sample Preparation

All single crystals of θ-RbZn, θ-CsZn and θ-TlCo were synthesized by the electrochemical method, as described in the literature [1]. The supporting electrolyte for θ-RbZn, θ-CsZn and θ-TlCo, is the mixture of MSCN (M = Rb, Cs, Tl), M'(SCN)$_2$ (M' = Zn, Co), 18-crown-6 ether in 1,1,2-trichloroethane, and 10% vol. of ethanol. The θ-RbZn and θ-CsZn were prepared by Dr. Miyagawa. The θ-TlCo was synthesized by Prof. Mori and Dr. Ueda at the University of Tokyo. The θ-RbZn and θ-CsZn single crystals have long and thin shape with the typical size of 0.1 mm × 0.2 mm × 1 mm. The θ-TlCo, on the other hand, has a block-like shape with the typical size of 0.2 mm × 0.5 mm × 0.5 mm.

2.2 Resistivity Measurements

The in-plane conductance of the series of materials was measured by using the conventional four-probe dc method. The gold wires of 15 or 25 μm φ were attached

© Springer Nature Singapore Pte. Ltd. 2017   23
T. Sato, *Transport and NMR Studies of Charge Glass in Organic Conductors with Quasi-triangular Lattices*, Springer Theses,
DOI 10.1007/978-981-10-5879-0_2

Fig. 2.1 Schematic picture
of electronic terminals

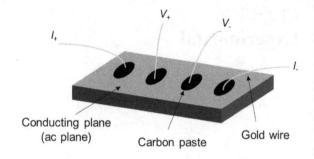

to the crystals with carbon paste, as shown in Fig. 2.1. The measurements were
performed under different levels of applied currents, and we took the data in the
ohmic regime as those free from the Joule heating.

2.3   Resistance Noise Measurements

In this thesis, we performed the resistance fluctuation spectroscopy, namely,
resistance noise measurements, for two different materials θ-CsZn and θ-RbZn. The
noise spectra were measured by the four-terminal dc method [2, 3]. A schematic
diagram of the noise measurements is shown in Fig. 2.2. The voltage drop under a
dc current along the c-axis generated by a low-noise voltage source was amplified
by a low-noise preamplifier (NF SA-400F3). Then generated voltage data in the
time domains were fed into a FFT analyzer (Keysight 35670A) to obtain the noise

Fig. 2.2 Schematic circuit
diagram for four-probe
resistance fluctuation
spectroscopy

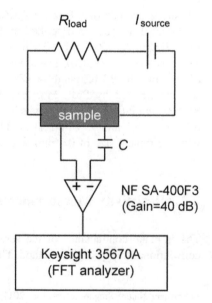

power spectrum density (PSD), S_R. The capacitor, C, is used as a high pass filter, which prevents the dc voltage offset from entering the FFT analyzer. Note that the resistance generally fluctuates in the time domains. Noise measurements capture the fluctuations in resistance as a form of voltage fluctuations by inducing a stable current.

Observed PSD inevitably includes three components as follows; (i) intrinsic fluctuations in sample resistance, (ii) background noise, and (iii) extrinsic noise, which mainly comes from the voltage source and/or contact noise at the sample, etc. Note that in our study, the non-essential noise of (iii) was successfully eliminated by using a large-load resistor, R_{load}, as confirmed below. Consequently, the observed PSD can be described in the following equation;

$$S_V(f, I) = G^2 \left[I^2 S_{sample}(f) + S_V^0(f) \right], \tag{2.1}$$

where G is the gain of the preamplifier. $S_V^0(f)$ is a floor noise, i.e. the background noise, $S_V(f, I = 0)$. It is $S_{sample}(f)$ that originates from intrinsic fluctuations in sample resistance [the term (i)]. In general, the sensitivity of the dc technique is governed by the background noise $S_V^0(f)$, which is dependent on the thermal noise and the preamplifier's noise figure (NF) defined as

$$\mathrm{NF} = 10\,\mathrm{dB} \times \left[\frac{S_V^0(f)}{S_{nyquist}} \right], \tag{2.2}$$

where $S_{nyquist}$ is the theoretical value of the thermal noise expected by the Nyquist rule, $S_{nyquist} = 4k_B T R$, in which k_B is Boltzmann's constant and T is the temperature measured [4]. Figure 2.3 shows the noise figure of the preamplifier (NF SA-400F3) we used, which gives a measure of the amount of noise added by the preamplifier above the thermal noise. By setting the measurements parameter closer to the "eye" of the noise figure, the preamplifier noise becomes lowered. The typical value of sample resistance we measured is in the order of 1–100 Ω, thus the impedance matching is almost achieved in our measurements.

Figure 2.4a displays the observed PSDs of θ-CsZn at 155 K taken at different bias currents I. We confirmed that the background thermal noise [black line in (a)] is frequency-independent (so-called "white noise") and nearly equal to the expected value expected by Eq. (2.2) (See the dashed horizontal line in Fig. 2.4a). It proves that extrinsic effects of (iii) were successfully eliminated. Then, the intrinsic resistance noise can be deduced by subtracting the value of thermal noise from that of S_V for $I \neq 0$, resulting in the 1/f-type PSDs as shown in Fig. 2.4b. Since one expects that the noise level of the voltage is proportional to the square of the applied current, $S_V \propto V^2 \propto I^2$, a necessary condition for checking if the measured noise indeed is solely due to resistance fluctuations of the sample is a scaling behavior above. The data in Fig. 2.5 is found to be scaled very well by the square of current. Also, as expected and required, the observed scaling is independent of the ratio

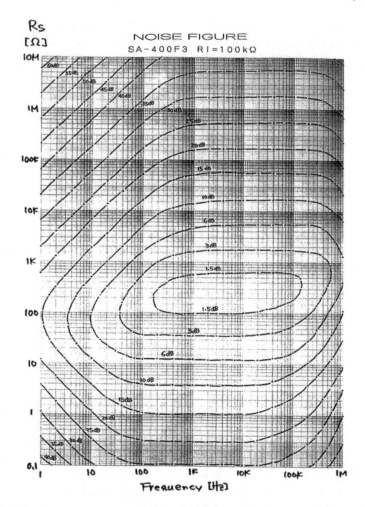

Fig. 2.3 Noise figure contour of NF SA-400F3 preamplifier

R_{load}/R, where R is the sample resistance (see Fig. 2.2). All these tests confirmed that the measured $1/f$ noise originates from the sample and is related to the resistance fluctuations in it.

2.4   X-Ray Diffuse Scattering Measurements

We also measured the X-ray diffraction of θ-CsZn, θ-RbZn, and θ-TlCo single crystal, using a Rigaku DSC imaging plate diffractometer and Si(111) monochromatized synchrotron radiation X-rays ($E = 18$ keV, $\lambda = 0.689$ Å) at the BL-8A and

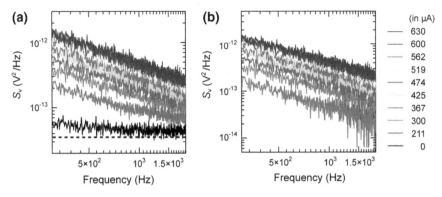

Fig. 2.4 Voltage noise PSD, S_V(V^2/Hz), for different bias currents I. **a** The observed raw data, and **b** deduced PSDs by subtracting the value of thermal noise from the value of S_V for $I \neq 0$ are shown, respectively. The data in (**b**) correspond to the intrinsic resistance noise

Fig. 2.5 Scaling behavior of S_V (integrated from 300 to 1900 Hz) against the square of bias currents, I. *Red* and *blue plots* represent the intensities of the spectra obtained for a different the load resistance

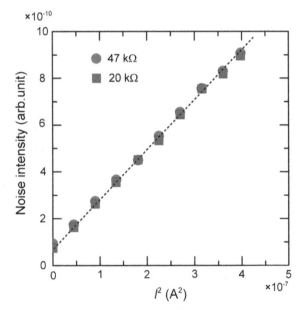

8B beamline of the Photon Factory (PF) at the High Energy Accelerator Research Organization (KEK). Since almost all the θ-TlCo are polycrystalline, we broke down them into smaller pieces with the size of 0.1 mm × 0.1 mm × 0.1 mm, being able to obtain clear oscillation photographs for single crystal. Each sample was mounted by Araldite on the tip of a glass pin, which is coated with Ag paste so as to conduct the heat efficiently (Fig. 2.6). Helium gas was sprayed on the samples to regulate the temperature, which makes it possible to cool them down with maximum rate of approximately 150 K/min.

Fig. 2.6 Schematic figure for
mounted sample for X-ray
scattering

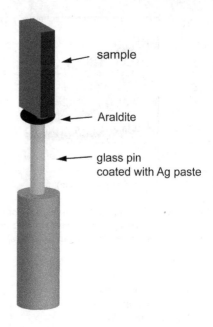

We analyzed the set of oscillation photographs of conducting planes, focusing
on the diffuse scatterings not on the Bragg reflections. The position and the line
profile of the observed diffuse scatterings provide the information about the wave
vector and the correlation length, ξ, of the short-ranged CO domains, respectively.
The ξ can be defined as the inverse of FWHM from the line profile, where FWHM
denotes the height of the full-width at half-maximum of the Lorentzian fit to the
data.

2.5 NMR Measurements

First, the block diagram of the homodyne-type NMR spectrometer used in this work
is shown in Fig. 2.7. Our experiments were performed under an external field (6 T).
The spin echo signals after the $(\pi/2)_x - \tau - (\pi)_x$ pulse sequence with $\tau = 10$ μs
were converted into NMR spectra through the Fast Fourier transformation. The
typical duration time of the $\pi/2$ pulse was 1.0 μs, which covers the NMR spectra
within approximately ± 80 kHz from the observation frequency. In each spectrum,
the line position of TMS (tetramethylsilane) is used as the origin of the NMR shift.

For the ^{13}C-NMR measurements, two central carbon atoms in the BEDT-TTF
molecules were selectively enriched by ^{13}C isotopes (Fig. 2.8). As mentioned in the
previous chapter, the crystal structure of θ-(BEDT-TTF)$_2$X at room temperature is
orthorhombic (with space group $I222$); thus, all BEDT-TTF molecules are crystallographically equivalent. As for the NMR spectra, however, each BEDT-TTF

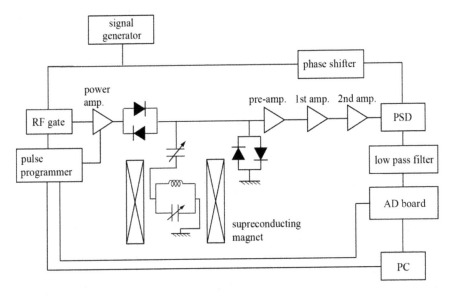

Fig. 2.7 Schematic diagram of the homodyne-type NMR spectrometer

Fig. 2.8 ^{13}C-substituted
BEDT-TTF molecule. *Arrows*
indicate the position of the
carbon atoms enriched by ^{13}C
isotopes

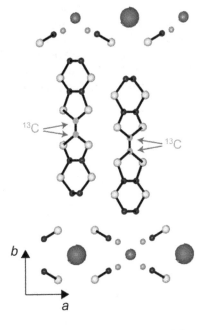

molecule with two ^{13}C nuclear spins gives a spectrum of multiple peaks up to 8 for
the following reasons (i)–(iii). (i) The two adjacent ^{13}C nuclei are non-equivalent in
the molecule and thus give separate resonance lines. (ii) The nuclear dipole interaction between the adjacent ^{13}C nuclei causes each resonance line split into a

doublet. The splitting width, d, depends on the angle, ϕ, between the directions of the applied magnetic field H and the $^{13}C = {}^{13}C$ vector (Fig. 2.9a), and is given by

$$d = \left(\frac{3}{2r^3}\right)\gamma_n^2\hbar\left(1 - 3\cos^2\phi\right), \qquad (2.3)$$

where $\gamma_n/2\pi = 10.7054$ MHz/T is the gyromagnetic ratio of the ^{13}C nucleus, $\hbar$ is the Planck constant divided by 2π, and r is the distance between the adjacent ^{13}C nuclei. (iii) BEDT-TTF molecules form a zig-zag configuration in the conducting layer, as shown in Fig. 2.9b. Therefore, the α and β molecules denoted in Fig. 2.9b are not equivalent with respect to the field directions, except for the particular ones described below, and exhibit separate spectra. Consequently, the number of resonance lines is a maximum total of $2 \times 2 \times 2 = 8$.

In our experiments, we applied the magnetic field in a particular direction to eliminate the splitting sources of (ii) and (iii). If the field direction is within the ab plane, then the α and β molecules are equivalent against the field direction and thus exhibit identical lines; namely, the third effect is eliminated. Furthermore, if the field is directed in the ab plane to form the "Magic angle" against the $^{13}C = {}^{13}C$ vector, ϕ, that satisfies the condition of $d = \left(\frac{3}{2r^3}\right)\gamma_n^2\hbar\left(1 - 3\cos^2\phi\right) = 0$, then the nuclear dipole coupling effect on spectra between the ^{13}C nuclei vanishes, leading to simply two lines positioned at the NMR shifts of the two ^{13}C nuclei. In the experiments, we first adjusted the magnetic field in a direction parallel to the b-axis and then rotated the field direction in the ab plane to realize the magic angle.

Figure 2.10 shows NMR spectra for θ-RbZn that were acquired under the field direction. At 295 K, the spectrum is comprised of two lines, which correspond to the two adjacent ^{13}C nuclei in a BEDT-TTF molecule (Fig. 2.10a). It confirms that

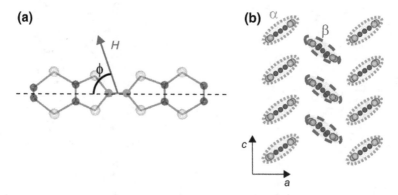

Fig. 2.9 a Definition of the angle, ϕ, between the applied magnetic field H and the line connecting adjacent ^{13}C nuclei. **b** *Top view* of the schematic arrangement of BEDT-TTF molecules in the conducting ac plane. Two molecules, α and β, (highlighted by *green dotted line* and *blue dashed line*, respectively) are nonequivalent with respect to arbitrary field direction except for the parallel field configurations to the ab or bc planes

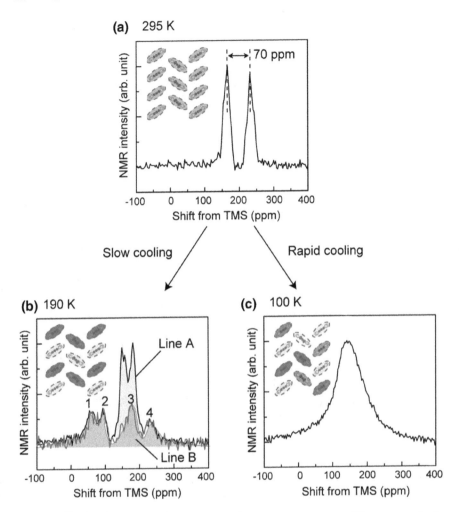

Fig. 2.10 ^{13}C-NMR spectrum for the θ-RbZn at several temperatures. **a** The spectrum in the metallic state at room temperature. The two peaks originate from the nonequivalent adjacent ^{13}C nuclei in the center of the molecule. **b** The spectrum in the CO state at 190 K when the sample was slowly cooled (at a rate of approximately 0.1 K/min) through T_{CO}. Spectra A and B are acquired in 1.5 s and 10 ms after the saturation of nuclear magnetization, respectively. **c** NMR spectrum after rapidly cooled (at a rate of 4 K/min) to 100 K

the applied field is precisely set in the desired angle described above, and the electronic state at room temperature is homogeneous.

When θ-RbZn is slowly cooled through T_{CO}, the spectrum changes its shape into a complicated structure that is characteristic of the CO state (Fig. 2.10b). The reconstructed spectrum below T_{CO} consists of two components: a doublet (line A, shaded in blue in Fig. 2.10b) and a broad line having four peaks (line B, shaded in red in Fig. 2.10b). Line B is observable at the time of 10 ms after the saturation of

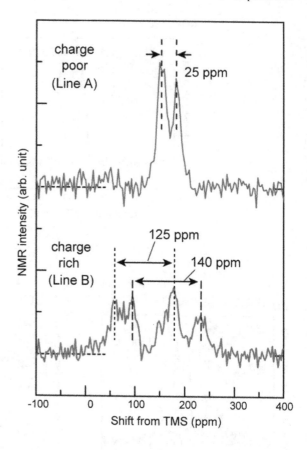

Fig. 2.11 Decomposed ^{13}C-NMR spectrum for the θ-RbZn at 190 K into charge-rich and charge-poor components

nuclear magnetization, whereas line A becomes appreciable only on the order of second or later after the saturation, indicating that the relaxation time of line A is far longer than that of line B. The relaxation time is inversely proportional to the square of the charge-density in the paramagnetic state; thus, the charge disproportionates in the CO state, in agreement with the previous result. Figure 2.11 shows line A, which is obtained by subtracting the 10-m sec spectrum from the 1.5-sec spectrum in Fig. 2.10b, as well as line B. Line A, originating from the charge poor molecule, is a doublet of a shift difference of $\delta_{poor} = 25$ ppm, whereas line B has a four peaked structure, although the separation of the left two lines is not so clear. The shift difference between the adjacent ^{13}C sites in a molecule is proportional to the local spin susceptibility on the molecule. Referring to the charge disproportionation ratio estimated at 295 K and the reported spin susceptibility, 6.1×10^{-4} emu/mol at 295 K and 7.1×10^{-4} emu/mol at 190 K [1], the shift differences in the poor and rich molecules are expected to be $\delta_{poor} = (7.1/6.1) \times (0.14/0.50) \times 70 = 23$ ppm and $\delta_{rich} = (7.1/6.1) \times (0.86/0.50) \times 70 = 140$ ppm. The profile of line A is consistent with this expectation. As for line B, its interpretation of the structure is not straightforward. Considering the expectation of $\delta_{rich} = 140$ ppm, it is likely that two nonequivalent charge-rich molecules with the sift differences of 140 and 125 ppm appears below T_{CO}, as indicated in Fig. 2.11. According to the X-ray study, all the charge-rich molecules are equivalent in crystallography [5]; as a result, the observed non-equivalence is attributable to the configuration of the molecule against the field direction. The X-ray study also revealed that the dihedral angle between the neighboring charge-rich molecules increases by 10° upon undergoing the charge ordering transition, i.e., the rotation of the molecules by 5° about the molecular long axis. In the present experiments, the possible error in fixing the field to the desired direction is estimated to be ±3°; this error may explain the subtle structure of each line at 295 K. It is likely that the small misalignment of magnetic field above T_{CO} is enlarged by the molecular rotation below T_{CO}, giving rise to the nonequivalence between α and β molecules in Fig. 2.9b. If the applied filed is rotated, e.g., by 3° in the ac-plane from magic angle, then the Knight shifts in molecules α and β are expected to differ by 35–40 ppm, referring to the known anisotropic hyperfine field. This quantitatively explains the emergence of the two pairs of lines (1–3 and 2–4), which originate from the charge-rich α molecules and the charge-rich β molecules.

Alternatively, when θ-RbZn is cooled rapidly across T_{CO}, the spectrum exhibits a broad and structureless feature (Fig. 2.10c), indicating a continuous distribution in charge density.

References

1. H. Mori, S. Tanaka, T. Mori, Phys. Rev. B **57**, 12023 (1998)
2. J. Müller, Chem. Phys. Chem. **12**, 1222 (2011)
3. J.H. Scofield, Rev. Sci. Instrum. **58**, 985 (1987)
4. H. Nyquist, Phys. Rev. **32**, 110 (1928)
5. M. Watanabe, Y. Noda, Y. Nogami, H. Mori, J. Phys. Soc. Jpn. **73**, 116 (2004)

Chapter 3
Charge-Glass State in θ-(BEDT-TTF)₂X

Abstract As pointed out in Chap. 1, the possible charge-glass behavior is expected to couple strongly to the degree of a geometrical frustration, namely, a triangularity of the lattice geometry. By changing anion X in the θ-(BEDT-TTF)$_2$X system, the degree of triangularity can be systematically tuned; therefore this family serve as an attractive platform to investigate glassiness under charge frustration. In this work, three different materials with different frustration parameter were studied; θ-(BEDT-TTF)$_2$CsZn(SCN)$_4$ (high frustrated), θ-(BEDT-TTF)$_2$RbZn(SCN)$_4$ (medium frustrated), and θ-(BEDT-TTF)$_2$TlCo(SCN)$_4$. In the beginning of this Chapter, focusing on the highest frustrated material in above three, θ-CsZn, which is suggested to be the most stable charge-glass former, we demonstrated the hallmarks of glass; non-equilibrium, slow dynamics, and short-range correlation. These findings clearly show that a charge-glass state is realized in organic conductor, θ-CsZn. It should be emphasized that the medium frustrated θ-RbZn and the least frustrated θ-TlCo are also revealed to be a charge-glass former, but the only difference lies in the critical cooling speed required for glass formation; a stronger frustration gives a higher-charge-glass former. This is the main result of the section. Our work also proposes that the cooling speed is potentially key control parameter to give rise to new electronic state.

Keywords θ-(BEDT-TTF)$_2$X · Charge order · Charge glass · Non-equilibrium · Slow dynamics · Short-range correlation · Fragility · Critical cooling rate · Glass-forming ability

3.1 Introduction

All of glass states would ubiquitously hold the pronounced features as follows; (i) non-equilibrium, (ii) slow dynamics, and (iii) short-ranged correlations in spatial domains. The glass state in charge degrees of freedom is also expected to possess these properties as the hallmarks of glass. This chapter describes the first experimental demonstrations of the charge-glass (CG) states in the organic conductor,

© Springer Nature Singapore Pte. Ltd. 2017
T. Sato, *Transport and NMR Studies of Charge Glass in Organic Conductors with Quasi-triangular Lattices*, Springer Theses, DOI 10.1007/978-981-10-5879-0_3

θ-(BEDT-TTF)$_2$X system. Distinctive from the case of structural glass, whose degrees of freedom are atoms or molecules, it is the electrons that bear the full nature of glass in CG system. Therefore, resistance, its noise and X-ray scattering can serve as a useful probe to capture the above hallmarks. We also examine the systematics of the emergent glassy phenomena in terms of the degree of the geometrical frustration to seek the correlation between the frustration and the charge-glass forming ability.

3.2   Hallmarks of Charge Glass in θ-(BEDT-TTF)$_2$CsZn (SCN)$_4$

Among the three different CG candidates, θ-CsZn, θ-RbZn, and θ-TlCo, we first focus on the θ-CsZn with the most isotropic triangular lattice, namely, the most highly charge frastrated, in which a CG state is expected to be the most stably formed. θ-CsZn is also advantageous for exploring the CG state over the whole temperature range owing to the absence of CO at least in the laboratory time scale. In the following subsections, we show the experimental manifestations of the three hallmarks of glass (i)–(iii) in turn.

3.2.1   Non-equilibrium Nature in Charge-Glass State

One of the hallmarks of a glassy state is that they fall out of thermodynamic equilibrium of a liquid (or supercooled) state, which occurs when the time scale of the fluctuations exceeds the time available for molecular relaxation allowed by the cooling rate. Therefore, the appearance of nonequilibrium nature strongly depends on the cooling speed [1]; the faster cooling leads to the higher experimental glass transition point, T_g. Thus, we first tested whether the charge vitrification depends on the cooling rate or not through electron transport measurements.

To reveal this signature, we revisited the ρ-T profile of θ-CsZn on cooling and heating at different temperature-sweep rates of 0.1, 1, 5, and 10 K/min. Appreciable hysteresises were observed at approximately 90–100 K (Fig. 3.1a), which is consistent with a previous report [2]. It was argued in Ref. [2] that this hysteresis is a signature of an inhomogeneously broadened first-order phase transition. Instead of such a viewpoint, we consider this hysteresis to be the manifestation of the formation of a charge-glass state. This notion is supported by the clear cooling-rate dependence of the thermal hysteresis as seen in Fig. 3.1a. They exhibit two noteworthy features. First, the hysteresis region depends on the cooling rate, Q. For clarity, we (tentatively) define T_g^*, which is determined from the ρ-T profile, as the temperature at which the difference between the resistivities in the heating and cooling processes takes a maximum. Figure 3.1b shows T_g^* versus Q, which shows

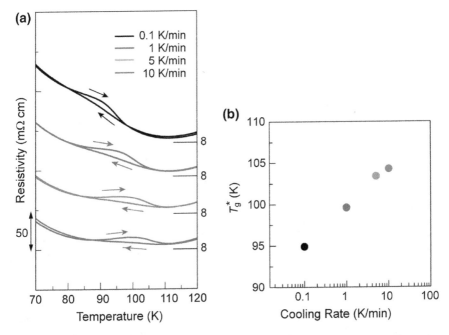

Fig. 3.1 a Temperature dependence of resistivity at various temperature-sweeping rates. **b** Experimental glass transition temperature versus temperature-sweeping rate

a positive correlation between them; therefore, the hysteresis around T_g^* must be kinetic rather than thermodynamic in origin. These experimental observations are in good agreement with the general glass behavior [3]. Second, the resistivity below T_g^* is appreciably dependent on Q. Figure 3.2 shows the temperature dependence of the resistivity normalized to the value at 120 K for different Q. It is clear that the normalize resistivity branches off, depending on the cooling rate, below approximately 100–110 K. Notably, more rapidly cooling leads to lower resistivity,

Fig. 3.2 Normalized resistivity profiles upon cooling at 0.1, 1, 5, and 10 K/min

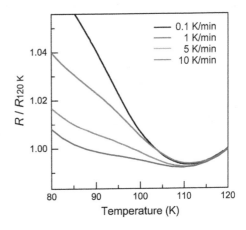

indicating that the charge configuration is frozen into a more unstable one. This behavior implies that the ground state would be an insulating CO state, although such transition has not been substantiated in the laboratory time scale. The deviation in resistivity provides further evidence of the glass transition upon cooe is reasonably close to T_g^* in Fig. 3.1b (~ 101 K for 10 K/min).

Now that the existence of T_g^* is made clear, it is reasonable to expect a nonequilibrium electronic state below T_g^*. In fact, this signature is already recognizable in Fig. 3.2: the low-temperature resistivity ($<T_g^* \sim 100$ K) is obviously Q-dependent. Even more compelling evidence for a nonequilibrium state is easily checked by the existence of "physical aging", which is known in the classical-glass system as the time-dependent structural rearrangement while keeping temperature [4]. It is the process of slow relaxation into a more stable configuration. In the present CG system, the physical aging can be recognized as the time-dependent resistivity variation after rapid cooling. For this purpose, the sample was first cooled down (~ 5 K/min) from high temperatures above T_g^* (~ 100 K) to a target temperature. Then, the time evolution of the resistance was monitored while the temperature was held fixed.

Figure 3.3 shows the time evolution of the resistivity at various temperatures. As expected, the aging behavior is clearly observed only in the temperature regime below T_g^*, indicating that the charge configurations falling out of thermodynamic equilibrium relaxed toward a more stable one with a very long relaxation time (e.g., up to several hours at 87.5 K). Notably, the characteristic time of the aging is longer at lower temperatures. For the quantitative evaluation, we fitted the set of aging curves by the well-known Kohlrausch-Williams-Watts (KWW) law, which is widely used to describe relaxation processes in supercooled liquids [5, 6]:

$$\rho(t) = \rho_0 + (\rho_\infty - \rho_0)\left[1 - \exp\left\{-\left(\frac{t}{\tau_{\text{aging}}}\right)^\beta\right\}\right], \qquad (3.1)$$

where ρ_0 and $\boldsymbol{\rho}_\infty$ denote the initial and final resistivity values in the aging process, respectively, which are obtained as fitting parameters. The τ_{aging} denotes the relaxation time and the β characterizes the nonexponentiality of the relaxation

Fig. 3.3 Aging behavior of resistance as a function of time at various temperatures. The *broken curves* are fits to the phenomenological relaxation behavior, Eq. (3.1)

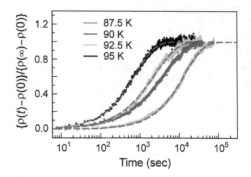

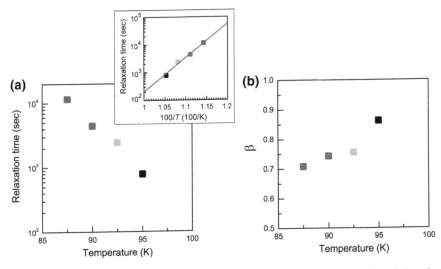

Fig. 3.4 **a** Temperature dependence of relaxation time deduced from the fitting of Eq. (3.1) to the resistance data. In the inset, the relaxation time is plotted in the Arrhenius form. **b** The stretched parameter, β, as a function of temperature

process. Figure 3.4a displays the temperature dependence of τ_{aging} obtained from the fitting of Eq. (3.1) to the data and they are plotted in the Arrhenius representation in the inset of Fig. 3.4a. As clearly shown in the inset, the data points of τ_{aging} are on a single Arrhenius function with a gap of 2600 K. Furthermore, just below the glass transition point, the value of β yields approximately 0.9, indicating a nearly single exponential relaxation (Fig. 3.4b). As temperature decreases from T_g^*, however, the β is decreased and approaches 0.7.

3.2.2 Slow Dynamics in Charge-Liquid State

The existence of the charge-glass transition can be further corroborated by the observation of its precursor, i.e., slow charge dynamics, in the thermodynamic equilibrium states above T_g^*. This is the second hallmark of glass as stated above. In the present case, such slow dynamics in charge above T_g^* is expected to be detectable as resistivity fluctuations in time domain [7]. Therefore, we performed the resistance fluctuation spectroscopy, i.e., noise measurements. The typical noise power spectral density (PSD) of the resistance fluctuations normalized by the square of resistance, S_R/R^2, is shown in Fig. 3.5a. Overall, the so-called 1/f noise is observed, that is, the PSD has $1/f^\alpha$-type frequency dependence with $\alpha = 0.8$–1.1, where f denotes frequency. It should be noted that the 1/f noise itself is ubiquitously seen in a wide range of systems, thus one usually considers that there is no characteristic time or frequency in the $1/f^\alpha$ noise with a particular α value. In the present

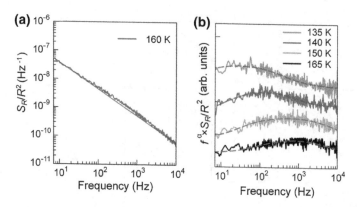

Fig. 3.5 **a** Typical power spectrum density of resistance fluctuations, S_R/R^2: the *straight line* is a fit of $1/f^\alpha$. **b** Power spectrum density multiplied by frequency at various temperatures: *broken curves* are fits to continuously distributed Lorentzians plus $1/f^\alpha$

case, however, there is observed an appreciable deviation from the background $1/f^\alpha$ fit (Fig. 3.5a). Focusing on this deviation, we extract the characteristic frequency f_c of this additional contribution to the fluctuations. For clarity, $f^\alpha \times S_R/R^2$ versus f is plotted in Fig. 3.5b; a broad but clear peak appears at f_c, which shifts towards lower frequencies with decreasing temperature.

To extract f_c, we introduce a hypothetical superposition of continuously distributed Lorentzians plus $1/f$ noise. In general, it is useful for the analysis of experimental PSD to refer to a two-level activation process, whose PSD has a Lorentzian type of function with a particular relaxation time. Within this scheme, $1/f$ noise is generated by the summation of a large number of two-level systems with widely distributed relaxation times. Based on this idea, we assumed the excess contribution to the $1/f$ noise in the limited frequency range between high-frequency f_{c1} and low-frequency f_{c2} cutoffs to result in the non-$1/f$ components. The fitting function we used is given by

$$S_R(f) = A \int_{\tau_{c1}}^{\tau_{c2}} G(\tau_c) \frac{\tau_c}{1 + (2\pi f \tau_c)^2} d\tau_c + \frac{B}{f^\alpha}, \tag{3.2}$$

where $\tau_{c1} = 1/2\pi f_{c1}$ and $\tau_{c2} = 1/2\pi f_{c2}$ correspond to high- and low-frequency cutoffs, respectively (Fig. 3.6). $G(\tau_c)$ represents the distribution function for the relaxation time, and is written by

$$G(\tau_c) \propto \begin{cases} 1/\tau_c & (\tau_{c1} \le \tau_c \le \tau_{c2}) \\ 0 & (otherwise) \end{cases}. \tag{3.3}$$

This formulation comes from the flat distribution function for the activation energy of two-level system, postulated for the occurrence of $1/f$ noise.

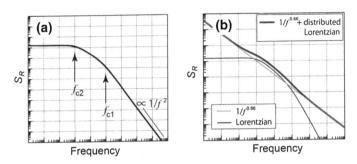

Fig. 3.6 a Simulated results of the power spectrum density S_R associated with the distributed Lorentzian, the first term in Eq. (3.2). **b** Power spectrum density when the $1/f$ noise is superposed on the distributed Lorentzian

$$G(\Delta E) \propto \begin{cases} const. & (\Delta E_1 \leq \Delta E \leq \Delta E_2) \\ 0 & (\text{otherwise}) \end{cases} . \tag{3.4}$$

Note that $G(\tau_c)$ is straightforwardly connected to $G(\Delta E)$, by the relationship of $\tau_c \propto \exp(\Delta E/k_B T)$ (Fig. 3.7). Although Eq. (3.2) is derived on the basis of some assumptions, the fitting of the data by Eq. (3.2) allows us to evaluate the temperature dependence of f_{c1} and f_{c2}, eventually giving an estimate of the characteristic frequency, f_c [$\equiv (f_{c1} f_{c2})^{1/2}$].

The fits of Eq. (3.2) to the data are shown by red broken curves in Fig. 3.5b. The fitting curves reproduce the spectra well, which enables us to extract the temperature dependence of f_c [$\equiv (f_{c1} f_{c2})^{1/2}$], as shown in Fig. 3.8. Here, a dramatic decrease in f_c over several orders of magnitude is observed upon cooling for two different samples (Fig. 3.8), clearly showing that the charge dynamics in the equilibrium regime slows down as temperature approaches the glass transition point (≈ 100 K). Moreover, as shown below, we confirmed that the relaxation time

Fig. 3.7 Assumed distribution function of an extra-contribution to PSD represented as a function of **a** activation energy or **b** relaxation time of the two-level systems

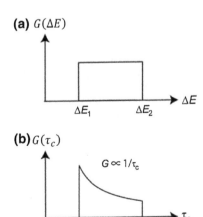

Fig. 3.8 Temperature
dependence of the
characteristic frequency, f_c,
$\left[\equiv (f_{c1}f_{c2})^{1/2}\right]$. The data are
extracted from the fits in
Fig. 3.5b

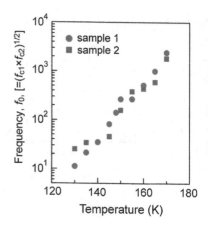

reaches 100–1000 s at $\approx$100 K, which is consistent with the conventional definition
of T_g^* (see Fig. 3.9).

Now, we have obtained the characteristic time scale of charge fluctuations for
temperatures both below and above T_g^*. Figure 3.9 displays the temperature dependence of τ_{aging} (see Fig. 3.4a) in the form of Arrhenius plot as well as the relaxation
times, τ_{noise} ($\equiv 1/2\pi f_c$), that were extracted using noise measurements above T_g^*
(Fig. 3.8). Remarkably, the temperature profiles of τ_{aging} and τ_{noise} can be well
described by the same equation, $\tau \propto \exp(-\Delta/k_B T)$ with the Δ/k_B value of
approximately 2600 K, strongly indicating the common charge dynamics for the
equilibrium states above T_g^* and the nonequilibrium states below T_g^*. It is plausible to
assume that the relevant charge dynamics consist of a rearrangement of the charge
configurations, which may be accompanied by a distortion of the local
lattice/molecules. Above T_g^*, thermodynamic equilibrium is achieved instanta-
neously owing to the short τ, and consequently, the charge fluctuations are centered
around the equilibrium states. Below T_g^*, in contrast, a longer time than a laboratory

Fig. 3.9 Temperature
dependence of the relaxation
time, τ_{aging}, derived from the
fits in (**a**) (*closed symbols*) as
well as the results shown in
Fig. 3.8 in units of seconds
(*open symbols*). The broken
line is the fit of the Arrhenius
function with a gap of
$\sim$2600 K

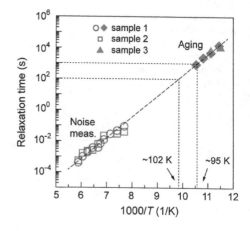

time scale is required to reach thermodynamic equilibrium, and thus a transient process from an initial to a (quasi)equilibrium final state is observed after the temperature is changed: this process is nothing but aging. Furthermore, the observed Arrhenius behavior tells us that the charge glass forming system θ-CsZn can obviously be classified as a "strong" glass, such as SiO$_2$. The strong glass generally implies that the glassy dynamics is dominated by an elementary process rather than a cooperative process; for instance, the local breaking and reforming of Si-O bonds are considered to play a major role in the glassy dynamics of SiO$_2$ [3, 8]. The strong-glass nature of θ-CsZn thus suggests that the rearrangement of charge configurations occurs locally. In the next section, we will discuss the nature of fluctuations and the possible origin of the strong-glass dynamics in detail by focusing on the global behavior of the 1/f noise.

3.2.3   Distribution of Activation Energy in Slow Charge Dynamics

In the previous section, we reported the results of noise measurements. They revealed that the resistance fluctuations of the 1/f-type with an appreciable deviation from that only above T_g, which is represented by a Lorentzian function. Focusing on this deviation, the characteristic frequency f_c related to charge fluctuations could be deduced, and its kinetic slowing down with an Arrhenius gap of approximately 2600 K was successfully captured. The uncovered nature of "strong glass" in θ-CsZn suggests that local fluctuations in charge rearrangement govern the glassy dynamics. However, the question of whether the local fluctuations are correlated or not, which is a keen interest in the physics of glass, remains experimentally unanswered.

Here, we revisit the observed PSDs, but now aim to investigate them by paying our full attention to characterizing the global behavior of the 1/f noise, especially the temperature dependence of noise magnitude in terms of the phenomenological model called Dutta-Dimon-Horn (DDH) model. This approach provides insight into the existence of cooperative fluctuations. Figure 3.10 displays several representative results of PSD for different temperatures. As we mentioned before, the overall behavior of S_R/R^2 is approximated as 1/f^α noise with $\alpha = 0.8$–1.1. If one looks closely at the frequency profile of S_R/R^2, however, it is seen that the observed PSD at a given temperature does not perfectly obey the 1/f^α-type frequency dependence with a single α value in the whole frequency range investigated. This frequency-dependent behavior of α is turned out to be well described by the DDH model, as explained below. To characterize the temperature dependence of S_R/R^2 and examine its frequency dependence, the values of γ and α, which corresponds to the magnitude and frequency exponent of 1/f noise respectively, are plotted as a function of temperature in Figs. 3.11a, b for 1 Hz and in Figs. 3.11c, d for 100 Hz. The

Fig. 3.10 Typical PSD of resistance fluctuations, S_R/R^2 at different temperatures

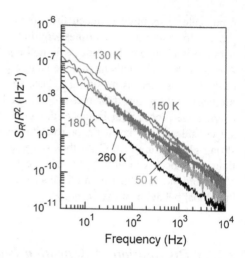

value of γ exhibits a characteristic temperature dependence with a maximum around 130 K for both frequencies. The γ and α values are used in the analysis below.

Here, we analyse the data in terms of the phenomenological model proposed by Dutta, Dimon, and Horn (DDH) [9], which is known to work well in a range of systems; inhomogeneous metals [9, 10], semiconductors [11, 12], superconductors in the normal state [13], and more recently ethylene glass in organic Mott insulators [7, 14]. The DDH model is based on the following two assumptions; (i) the $1/f$ noise is generated by the superposition of thermally activated two-level processes (called fluctuators), which couple to the resistivity in the present case, and (ii) each fluctuator is uncorrelated.

The relaxation time of the fluctuator, τ, is given by

$$\tau = \tau_0 \exp(E/k_B T), \tag{3.5}$$

where τ_0 is an attempt time very likely related to phonon frequency (typically 10^{-12}–10^{-14} s), E is the activation energy, which is different from fluctuator to fluctuator, and k_B is Boltzmann's constant. The PSD of a single fluctuator is a Lorentzian according to the Wiener–Khintchine theorem [15]. The noise spectrum is given by the summation of the Lorentzian spectra with various τ over fluctuators, and written by

$$S(f) \propto \int g(T) \frac{\tau(E)}{\tau(E)^2 4\pi^2 f^2 + 1} D(E) dE, \tag{3.6}$$

where $D(E)$ is the distribution of activation energies, and $g(T)$ is the temperature-dependent weighting function, which represents the change in the number of the fluctuators or coupling between fluctuators and resistance as a function of temperature. If $D(E)$ varies slowly in energy compared with $k_B T$, the distribution of

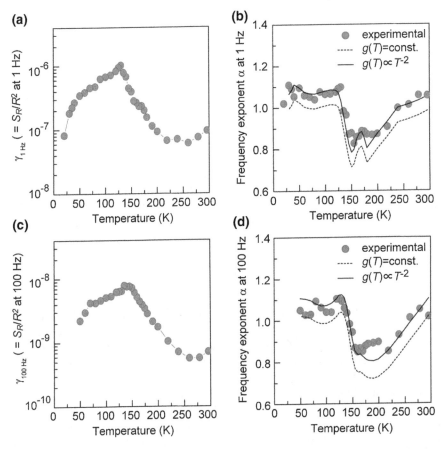

Fig. 3.11 a and **c** Temperature dependence of the magnitude of the power spectrum density, $\gamma_{1Hz}(= S_R/R^2$ at 1 Hz) and $\gamma_{100Hz}(= S_R/R^2$ at 100 Hz). **b** and **d** Temperature dependence of the frequency exponent α at 1 and 100 Hz. Solid and dashed lines trace the theoretical values of α_{DDH} deduced from the temperature dependence of S_R/R^2 in (**a**) by using Eq. (3.9) with $g(T) = \textbf{\textit{const.}}$ and $g(T) \propto T^{-2}$, respectively

energy may be deduced directly from the temperature dependence of the noise spectrum by

$$D(E) \propto \frac{2\pi f S(f, T)}{k_B T} \frac{1}{g(T)}. \tag{3.7}$$

Note that $S(f, T)$ at given f and T values is mainly contributed to by a fluctuator with the actication energy,

$$E \approx -k_B T \ln(2\pi f \tau_0), \tag{3.8}$$

as deduced from Eq. (3.5) and the Lorentzian form in Eq. (3.6). The present experimental range of 1 Hz $< f < 10^5$ Hz and 20 K $< T <$ 300 K covers the energy scale of 0.05–0.8 eV. To obtain $D(E)$ is the main goal of DDH analysis.

If $D(E)$ is constant, γ [$\propto S(f,T)$] is only weakly temperature-dependent according to Eq. (3.7). However, we observed the strong temperature dependences peaked around 130 K (Figs. 3.11a, c), suggesting that $D(E)$ is noticeably E-dependent. If $D(E)$ has a particular energy gap of E^*, namely $D(E) = \delta(E - E^*)$, as the opposite case, γ has a sharp peak structure at the temperature, T^*, which is linked to E^* through Eq. (3.8), with exponential temperature dependences above and below T^*. As the T dependences of γ in Figs. 3.11a, c are not so steep as the exponential, $D(E)$ is likely a broadened function with a peak around 0.3 eV according to Eq. (3.8). The concrete form of $D(E)$ is obtained by means of the DDH analysis below.

In the DDH model, the frequency exponent α_{DDH} is expressed by using Eqs. (3.7) and (3.8) as

$$\alpha_{DDH}(T) = 1 - \frac{1}{\ln(2\pi f \tau_0)} \left[\frac{\partial \ln S(f,T)}{\partial \ln T} - \frac{\partial \ln g(T)}{\partial \ln T} - 1 \right]. \qquad (3.9)$$

Thus, whether the DDH model is relevant to the experiments can be tested by comparing the observed $\alpha(T)$ with the $\alpha_{DDH}(T)$ given by Eq. (3.9). Note that α is frequency-dependent; so, $\alpha(T)$ and $\alpha_{DDH}(T)$ should be evaluated in the similar frequency range. Below, we compare the two values in the two frequency ranges around 1 and 100 Hz.

The predicted $\alpha_{DDH}(T)$, in which we used S_R/R^2 at $f = 1$ and 100 Hz, are shown as well as the observed $\alpha(T)$ in Figs. 3.11b, d, respectively. We used the attempt time $\tau_0 = 10^{-13}$ s, to which the results described below is extremely insensitive because of the logarithmic dependence of the energy on frequency and attempt time. The α_{DDH} with $g(T) = const.$ appears to be shifted by a constant from the experimental $\alpha(T)$ irrespectively of temperature, indicating that $g(T)$ is in a form of a power of T, $g(T) \propto T^b$. Indeed, the exponent of $b = -2 \pm 0.3$ well fits $\alpha_{DDH}(T)$ to the experimental $\alpha(T)$, indicating that the resistance noise in the present CG is describable by the assembly of uncorrelated fluctuators [the assumptions (i) and (ii)]. Then, $S_R(T)$ is transformed into $D(E)$, considering that T and E are connected through Eq. (3.8). The $S_R(T)$ at different frequency (1 and 100 Hz) gives the almost same results; the plot of $D(E)g(T)$ versus E shows a shoulder-like structure around 0.3 eV (Fig. 3.12), which is more visible in the plot of $D(E)$ versus E. The structure of $D(E)$ has a distinct peak at around 0.3 eV (Fig. 3.13). Notably, the value of 0.3 eV is of the order of the previously reported Arrhenius gap ($\Delta E \approx$ 2600 K) deduced from the analysis of the Lorentzian components superposed on the $1/f$ noise.

The present results indicate that the particular energy of 0.3 eV dominates the charge fluctuations in the present CG system. Furthermore, the agreement of the experimental α values with the values expected from the DDH model suggests that

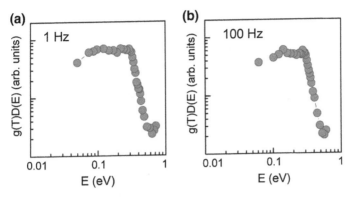

Fig. 3.12 E dependence of $g(T)D(E)$, calcurated from the data of **a** 1 Hz and **b** 100 Hz. Each plot is deduced from the data in Figs. 3.11a, c, respectively

Fig. 3.13 Distribution function of activation energy of the fluctuator, $D(E)$, calcurated from the data in Figs. 3.11a, c [assuming $g(T) \propto T^{-2}$]. The filled and open circles indicate the data of 1 and 100 Hz, respectively

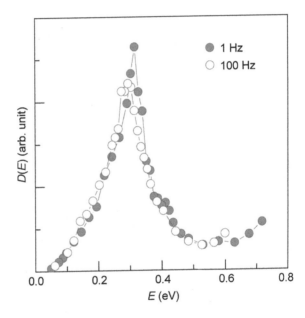

the fluctuations occur randomly and independently in a temperature range across T_g. In other words, the fluctuations are uncorrelated. These findings make complete sense with the strong-glass nature of θ-CsZn.

Considering the spatial inhomogeneity of the glassy state to be explained in the next section, the fluctuators should have a short correlation length throughout the temperatures investigated. It is noted that the characteristic energy of charge fluctuations, 0.3 eV, is comparable to the inter-site Coulomb repulsion, $V = 0.1$–0.3 eV, which is suggested from various experiments [16, 17]. This implies that the rearrangement in charge configuration is due to one electron hopping to neighbour sites with the activation energy of V. We suppose that the DDH model works well owing to the local nature of charge dynamics.

3.2.4 Evolution of Short-Ranged Charge-Ordered Domains

Finally, we demonstrate the evolution of the short-ranged correlation in the spatial domains, which is a hallmark of glass. Some simulation results for supercooled normal liquids argued that the formation of short-range or even medium-range clusters is a critical issue for understanding the slow dynamics at a microscopic level [18–21]. To see whether the previously reported short-ranged CO (or charge cluster) is correlated with the development of the slow dynamics described above [22, 23], we conducted x-ray diffuse scattering measurements at a synchrotron facility. In line with the previous observations [22, 23], two different diffuse scattering characterized by the wave vectors $q_1 \sim (2/3, k, 1/3)$ and $q_2 \sim (0, k, 1/2)$ were found near the Bragg spots (Fig. 3.14) (k is meant to express negligible coherence between the BEDT-TTF layers), that is, "3 × 3"-period and "1 × 2"-period CO domains are present.

Here, we focus on how short-ranged CO domains grow up at lower temperatures. Figure 3.15a shows the temperature dependence of the line profile along a^* axis for the diffuse scattering around q_1 vector. As temperature decreases, the linewidth of the q_1 CO domains narrows, indicating that the cluster size increases (Fig. 3.15a). The q_2 CO domains also show the similar tendency. The size of the CO domains, ξ, is estimated from the inverse of the half width at half maximum (HWHM) of the line profiles, which was fitted by the Lorentzian-type function.

The ξ-T profile is shown in Fig. 3.15b, where the size of the "3 × 3"-period (q_1) CO domains grows as the temperature decreases but levels off below approximately 100 K; that is, the "3 × 3"-period charge clusters appear to be frozen in no excess of 80 Å. This behavior is distinct from the conventional critical phenomena, where ξ diverges for a continuous transition. Furthermore, the temperature at which the cluster size ceases to develop is near the T_g value that was determined from the

Fig. 3.14 Oscillation photograph of the $a^* - c^*$ plane at 33 K. Diffuse scattering patterns characterized by $q_1 \sim (2/3, k, 1/3)$ and $q_2 \sim (0, k, 1/2)$ can be observed

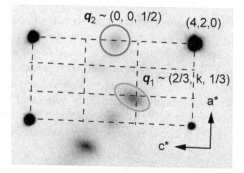

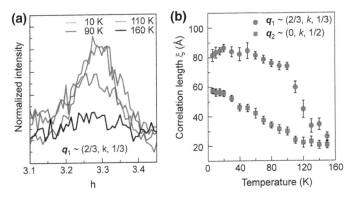

Fig. 3.15 **a** Line profile of the q_1 diffuse scattering along the a^* axis at various temperatures. **b** Temperature dependence of the cluster sizes of q_1 (*circles*) and q_2 (*squares*) charge clusters

transport measurements (Figs. 3.1b and 3.9), indicating that the spatial growth of the "3 × 3"-period CO domains is strongly coupled with the underlying charge dynamics. Interestingly, the "1 × 2"-period (q_2) CO domains continue to grow with decreasing temperatures even below T_g (≈100 K), suggesting that the charge dynamics associated with the "1 × 2"-period charge clusters are still active. Conversely, there appears to be "free space" for the "1 × 2"-period charge clusters to grow. Based on these observations, we conjecture that at approximately T_g^*, the "3 × 3"-period CO domains are frozen sparsely in the sea of the "para" state, thereby allowing the subsequent growth of another type of CO domains at lower temperatures.

In conclusion, we have demonstrated the emergence of CG state in θ-CsZn. The nonequilibrium nature of the charge dynamics in the CG state was evidenced by the cooling-rate-dependent charge vitrification and physical aging of the resistance. Moreover, noise measurements clearly showed that the temperature dependence of the relaxation time follows the Arrhenius law, suggesting that the glassy charge dynamics are caused by the local rearrangement of charge configurations. This is further confirmed by the DDH analysis of the noise spectra. X-ray diffuse scattering measurements revealed that the spatial growth of the "3 × 3"-period CO domains is closely related to the glassy charge dynamics. All of these experimental observations show that θ-CsZn is unequivocally CG formers.

Finally, we briefly discuss the origin of the strong-glass nature. The strong glass generally requires the local rearrangement motion in charge, as mentioned above. This scenario is seemingly available only when the charge configuration after one electron hopping is also metastable. Such a situation is more likely under stronger geometrical frustration, which causes a larger number of locally different charge configurations of similar energies, resulting in less cooperative dynamics. Therefore, we consider that the strong-glass nature in θ-CsZn is the consequence of strong geometrical frustration. Natural questions then emerge: what is the fragility of the more weakly frustrated systems, θ-RbZn and θ-TlCo?—are they fragile-glasses? Furthermore, is the geometrical frustration the dominant factor in causing the glassy freezing? In the following section, we compare the properties of θ-RbZn, and θ-TlCo with those of θ-CsZn, shedding light on the above questions. Investigating the material dependence is expected to bring about the comprehensive understanding of the CG state, hopefully offering a clue to the generic problem in physics of glass, namely, the role of frustration in the glassy freezing.

3.3 The Effect of Geometrical Frustration on the Glassy Behaviors

As described in Sect. 3.2, we demonstrated that θ-CsZn is a CG former and also pointed out the importance of charge frustration in forming the CG state. To look further into the effect of frustration, we investigated the θ-RbZn with a more weakly frustrated (more anisotropic) lattice than that of θ-CsZn, shedding light on the precise correlation between charge frustration and glassy behaviors.

3.3.1 Charge-Glass State and Its Fragility in θ-(BEDT-TTF)$_2$RbZn(SCN)$_4$

As we explained in Chap. 1, the θ-RbZn has the distinct cooling-rate dependence in the temperature-resistivity $(T - \rho)$ profile; slow cooling causes a transition to the conventional CO state $(T_{CO} = 198\ K)$, whereas rapid cooling preserves the high-conducting state that is continuously connected to the charge-liquid state above T_{CO} (Fig. 1.6). According to the previous NMR spectra, the electronic state of the rapidly cooled θ-RbZn is quite similar to that of θ-CsZn, strongly indicative of the emergent CG state in θ-RbZn [24, 25]. However, the hallmarks of glass have yet to be revealed for θ-RbZn. It is mainly because the putative CG state in θ-RbZn possesses a strong tendency towards a long-ranged CO state in the wide temperature range below T_{CO}, preventing one from investigating nonequilibrium phenomena in θ-RbZn while maintaining the glass state; for example, a relaxation towards the long-ranged CO state inevitably overwhelms possible physical aging in the glass state. We then seek the slow charge dynamics related to glass in charge-liquid state above T_{CO}, and therefore performed the noise measurements to capture it as we did for θ-CsZn.

Figure 3.16a displays the power spectrum density, S_R, divided by the square of resistance R^2 observed at 235 K. As expected, an appreciable deviation from the background $1/f$ noise over a certain frequency range is successfully observed. To quantify the non-$1/f$ contributions, we applied the same analytical procedure to the present data as for the θ-CsZn. In Fig. 3.16b, $f^\alpha \times S_R/R^2$ is plotted against f for several temperatures, where additional contributions appear as broad peaks over a

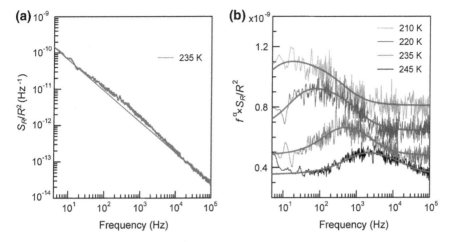

Fig. 3.16 Resistance fluctuations in the charge-liquid phase. **a** A typical resistance power spectrum density S_R normalized by the squared resistance R^2. **b** Power spectrum densities for various temperatures in the $f^\alpha \times S_R/R^2$ representation. The *red lines* are the fits of the distributed Lorentzian model

constant background. The peak position shifts toward lower frequency as temperature is lowered. The red lines in Fig. 3.16b show the fitting curves of the distributed Lorentzian model [Eq. (3.2)], which reproduce the measured spectra very well over the frequency range we measured. The obtained high-frequency and low-frequency cutoffs, f_{c1} and f_{c2}, give estimates of the characteristic frequency in charge fluctuations, f_c [$\equiv (f_{c1} f_{c2})^{1/2}$]. The resulting f_c values are plotted against temperature in Fig. 3.17. Clearly, f_c slows down by a couple of orders of magnitude as the temperature decreases; remarkably, just above the charge-ordered transition (approximately 200 K), the lifetime of the fluctuator is of the order of 10 Hz (Fig. 3.17). Thus, we conclude that the slow charge dynamics is present in the high-temperature charge-liquid state in θ-RbZn as well.

We also plotted the characteristic relaxation time in the Arrhenius form as shown in Fig. 3.18. The data follow the Arrhenius-type of function with the gap of approximately 5600 K. Interestingly, the gap value is roughly twice larger than that for θ-CsZn ($\sim$2600 K). Although an obvious deviation from the strong-glass behavior, e.g., super-Arrhenius behavior, is not found, the larger gap value imply that fluctuators may be mutually interacting, or some collective fluctuations may occur in θ-RbZn. To compare θ-RbZn with θ-CsZn in fragility, we display the so-called "Angell plot" that is, the Arrhenius plot of the relaxation time scaled by T_g (Fig. 3.19) [1, 26], where the relaxation time in θ-RbZn is assumed to obey the Arrhenius law even below T_{CO}, giving the empirically defined T_g value of approximately 160 K as the point at which the relaxation time reaches 100 s. The

Fig. 3.17 Temperature
dependence of the
characteristic frequency, f_c

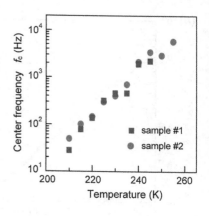

Fig. 3.18 Temperature
dependence of relaxation
time, which is derived from f_c,
in the form of Arrhenius
representation. The *broken
line* indicates a fitting line
with the gap of 5600 K. The
dotted lines are guides for the
estimation of the empirical T_g,
at which the relaxation time is
expected to reach 100–1000 s

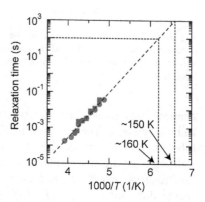

Fig. 3.19 Angell plot for the
θ-CsZn and θ-RbZn

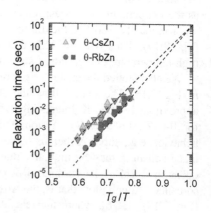

degree of fragility is generally assessed from the slope of the plot near T_g, which is called the fragility index, m; namely [6]

$$m = \left. \frac{d(\log \tau)}{d\left(T_g/T\right)} \right|_{T=T_g}. \tag{3.10}$$

The larger value of m corresponds to the steeper slope in Angell plot, that is, more fragile.

The broken lines in Fig. 3.19 yields the m values of 35 for θ-RbZn and 28 for θ-CsZn. Both values are much smaller than the typical m values of fragile glass (>100 [27]), being not inconsistent with the assumption of Arrhenius-type form in slowing down. It is noted, however, that θ-RbZn is more fragile than θ-CsZn. The increased fragility in θ-RbZn is possibly due to the reduction of geometrical frustration in line with the above discussions. This tendency is consistent with the theoretical suggestions in classical glass [21], thus implying the universal relation between the geometrical frustration and the nature of glass.

3.3.2   Evolution of Short-Ranged Spatial Order in θ-(BEDT-TTF)₂RbZn(SCN)₄

In θ-CsZn, two different diffuse scatterings characterized by the wave vectors $q_1 \sim (2/3, k, 1/3)$ and $q_2 \sim (0, k, 1/2)$ were found and revealed to be connected with the CG transition. To see the spatial correlation and its relation to the observed slow dynamics in θ-RbZn with the larger gap, we conducted X-ray diffuse scattering measurements for θ-RbZn.

A typical oscillation photograph for θ-RbZn is shown in Fig. 3.20a. As previously reported [28], a diffuse scattering with $q_1 \sim (1/3, k, 1/4)$ is observed, where the notation of means the absence of a well-defined wave vector, namely, quite a weak coherence between the conducting layers; in other words, the correlation is nearly confined in two dimensions. Note that the 3×4 periodicity in θ-RbZn, is different from the 3×3 periodicity in θ-CsZn. Remarkably, there exists no nucleus of the horizontal CO with $q_2[=(0, 0, 1/2)]$ above 200 K.

The temperature profile of the CO domains can be found in Figs. 3.20b and 3.21a, which show the evolution of the intensity and the correlation length ξ as a function of temperature. The slight asymmetry in the line profile is due to the background asymmetry, which is temperature independent and irrelevant to the temperature-dependent diffuse scattering (Fig. 3.20b). Since the q_1 modulation is replaced by the q_2 one below 200 K owing to the emergence of the twofold horizontal CO through the first-order transition on slow cooling, the temperature range of measurements under slow cooling is limited to above 200 K. The resultant value of ξ is, for example, ~ 140 Å at 210 K, which corresponds to ~ 25 triangular lattices. Compared between Figs. 3.21a and 3.17, the growths of the slow dynamics

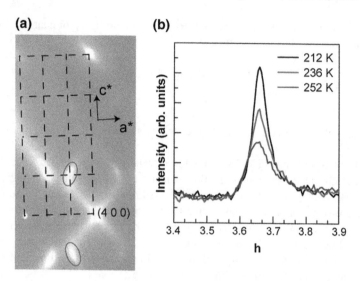

Fig. 3.20 a Photographic oscillation image of the $a^* - c^*$ plane at 225 K. Diffuse rods characterized by $q_1 \sim (1/3, k, 1/4)$ are observed near the Bragg reflections. **b** Line profile of $q_1 \sim (1/3, k, 1/4)$ along the $-2a^* + c^*$ direction

and of ξ appear well correlated; the CO domains with 3×4-period periodicity, not 1×2-period, are strongly coupled to the slow dynamics, in line with the consequence from the results for θ-CsZn. All of the physical properties discussed so far are interpreted as the precursory manifestation of the charge glass frozen below 200 K by rapid cooling. Thus, the last issue to be studied is the CG transition, which is expected to occur only if the frustration-relaxing transition at 200 K is prevented by rapid cooling. We measured the diffuse scatterings during heating after rapid cooling (~ 90 K/min; Fig. 3.21b) to see the temperature-correlation

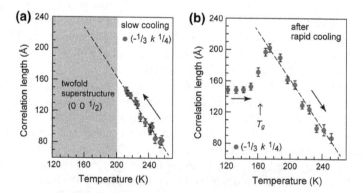

Fig. 3.21 Temperature dependence of the correlation length of CO domains, ξ, **a** during slow cooling and **b** during heating after rapid cooling to 120 K. The broken lines in (**a**) and (**b**) are guides for the eyes. The *error bars* represent the ambiguity of the fitting

length (T-ξ) profile. After rapid cooling to 120 K, only the diffuse scattering of $q_1 \sim (1/3, k, 1/4)$ is observed. During heating from 120 to 150 K, ξ of the q_1 reflection keeps a value of 145 Å, which is significantly shorter than expected from the extrapolation of the high-temperature behavior (see also Fig. 3.21a), and no superlattice reflections such as the $q_2[=(0, 0, 1/2)]$ reflection are observed within the measurement time (of the order of hours). This frozen state with no long-ranged order is characteristic of a glassy state in line with the behavior of θ-CsZn. Upon further heating, however, the value of ξ increases sharply around 160–170 K, reaching 200 Å at 175 K, and then turns to a decrease, which smoothly continues to the initial behaviour above 200 K before the rapid cooling. This behavior demonstrates the melting of the CG around 160–170 K.

The difference between the charge-cluster domain sizes of θ-RbZn and θ-CsZn in the glass state is noteworthy. The cluster size is 150 Å in θ-RbZn (3×4-periodicity) and ~ 80 Å for θ-CsZn (3×3-periodicity). The larger spatial correlation length may also be due to the increasing fragility. Thus, we see a reasonable relation between charge-cluster size, fragility and the geometrical frustration.

3.4 The Effect of Geometrical Frustration on Charge-Glass-Forming Ability

In the previous section, we argued the possible correlation between the charge frustration and the fragility of the CG state by comparing the results for θ-CsZn and θ-RbZn. In this section, we aim to comprehend the CG behavior of the θ-(BEDT-TTF)$_2$X family in terms of "charge-glass forming ability". The systematics of the CG behavior indicates a clear tendency that charge frustration increases the charge-glass forming ability in accordance with theoretical predictions in the general glass physics.

3.4.1 Stability of the Charge-Ordered State in θ-(BEDT-TTF)$_2$TlCo(SCN)$_4$

As shown in Fig. 1.6, all of the three compounds under investigation exhibit a low-resistivity state at high temperatures, which is referred to as the charge-liquid state in this study. In θ-RbZn, the charge-liquid state, while supercooled and followed by the CG formation by rapid cooling at a rate faster than 5 K/min, gives way to a horizontal CO under slow cooling. In θ-CsZn, the charge liquid state does not undergo a transition to the long-ranged CO even under slow cooling at a rate of as low as 0.1 K/min but leads to the CG formation below 100 K in a continuous manner. The triangular lattice of θ-TlCo is more anisotropic than both of θ-RbZn and θ-CsZn, thus offering an example as the least frustrated system. However, less

Fig. 3.22 Temperature
dependence of the resistivity
of θ-TlCo measured on
cooling at the rates of 0.3 and
10 K/min. Inset shows the
temperature dependence of
the resistivity of θ-RbZn
measured under cooling at the
rates of 0.1 and 4 K/min

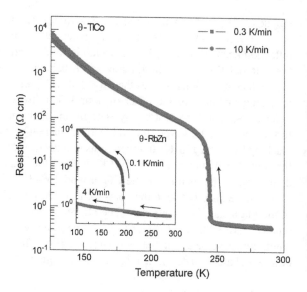

has been known for θ-TlCo; the only work reported is the optical study by Suzuki et al., who determined the stripe pattern of CO in the insulating state [29].

Here, we report transport and X-ray studies performed under slow and rapid cooling conditions for θ-TlCo. Figure 3.22 shows the temperature dependence of the resistivity measured when slowly cooled at a rate of 0.3 K/min. As reported in the literature [30], a sharp first-order transition from the low-resistivity charge-liquid state to the high-resistivity CO state is observed at 245 K. We also measured resistivity on faster cooling; however, no appreciable difference was detected even though the cooling rate is raised to 10 K/min, which is sufficiently fast to kinetically suppress the charge ordering in θ–RbZn (see the inset of Fig. 3.22), indicating that the CO state is more stable in θ-TlCo than in θ–RbZn. X-ray diffraction measurements further verified the robustness of the CO state. The sample was cooled at a rate of 150 K/min across the transition temperature (T_{CO} = 245 K); nevertheless, a superlattice reflection characterized by an in-plane wave vector, $q_2 = (0, 0, 1/2)$ (i.e., a horizontal CO), is clearly observed below T_{CO} in the oscillation photograph (Fig. 3.23a).

The first-order CO transition in θ-TlCo was not suppressed kinetically in the present study; however, we observed a symptom of charge vitrification in the X-ray diffuse scatterings in the charge-liquid phase. A typical oscillation photograph of θ-TlCo in the charge-liquid phase is shown in Fig. 3.23b, where faint diffuse spots with an in-plane wave vector $q_1 = (1/3, k, 1/6)$ (i.e., a signature of CO domains) are recognizable, being the possible manifestation of the incipient charge-glass formation. It is remarkable that the wave vector of the CO domain differs among the three systems (see also Table 3.1). Although we have no convincing explanation of this variation, which will invoke a future theoretical study [31], the diversity of the modulation wave vectors suggests that various charge configurations are competing

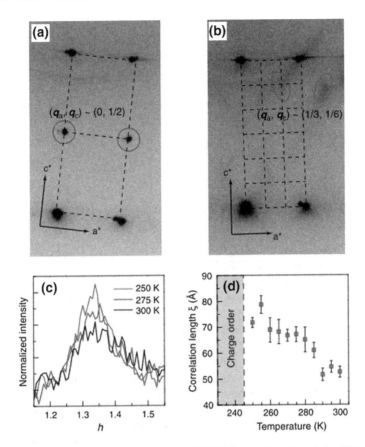

Fig. 3.23 Oscillation photograph of θ-TlCo **a** at 100 K (after passing through the CO transition temperature of 245 K at a quenching rate of 150 K/min) and **b** at 300 K. **c** Line profile of the diffuse scattering along the a^*-axis at various temperatures. **d** Temperature dependence of the size of CO domains in the charge-liquid phase

Table 3.1 Summary of the physical properties of the θ-(BEDT-TTF)$_2$X systems

	θ-CsZn	θ-RbZn	θ-TlCo
Charge frustration, V_p/V_c	0.92	0.87	0.84
T_{co} under 0.1–0.3 K/min cooling	N/A	~195	~245
Wave vector (q_a, q_c) of CO domains	(2/3, 1/3)	(1/3, 1/4)	(1/3, 1/6)
Critical cooling rate, R_c (K/min)	<0.1	1–5	>150
CG forming ability	High	Medium	Low

with a delicate balance under geometrical frustration. In the cases of θ-RbZn and θ-CsZn, the profile of the X-ray diffuse scatterings evolved in both of intensity and correlation length on cooling, which is interpreted as a signature of systems trending toward the CG transition. The line profile of the diffuse scattering in θ-TlCo also exhibits an appreciable temperature dependence in 250–300 K (Fig. 3.23c).

Figure 3.23d shows the correlation length of the diffuse scattering or the size of the "3 × 6"-period CO domains, ξ, estimated from the inverse of the half width at half maximum of the line profiles along a^*-axis. Compared with the case of θ-RbZn, the temperature evolution of the cluster size in θ-TlCo is moderate. This can be a characteristic of the system with less trend toward charge-glass. The cluster size just before the CO transition appears to show a meaningful difference between them; $\xi \sim 70$ Å for θ-TlCo and ~ 150 Å for θ-RbZn.

3.4.2 Systematic Variation of the Charge-Glass-Forming Ability

As see above, the θ-(BEDT-TTF)$_2$X systems exhibit diverse physical properties in not only the thermodynamic equilibrium states but also the nonequilibrium glassy states. For the comprehensive understanding of them, we here introduce the concept of "charge-glass-forming ability". It follows the notion of the glass-forming ability that is used in the physics of oxide glasses [32] and metallic glasses [33–35]. The (charge-)glass-forming ability is closely associated with the speed of the (charge-) crystallization from the glass state or supercooled liquid and can be evaluated quantitatively from the critical cooling rate, R_c, that is the minimum cooling rate required for the kinetic avoidance of the first-order (charge-)crystallization. A lower R_c signifies slower kinetics toward the CO ordering and hence points to a higher (charge-)glass-forming ability. Below, we estimate the values of R_c (or charge-glass-forming ability) experimentally for the present three materials, θ-RbZn, θ-CsZn, and θ-TlCo.

In θ-RbZn, the charge crystallization, which occurs on cooling at a rate less than 1 K/min, is suppressed when cooled at a rate of 5 K/min, indicating that the R_c value is in between 1 and 5 K/min. In θ-CsZn, the CO transition is absent at least in the laboratory time scale. However, considering that the CG is a nonequilibrium state toward the CO, one can postulate that the CG state in θ-CsZn results from the kinetic avoidance of some type of CO state; that is, it may be a case that R_c of θ-CsZn is finite but beyond the laboratory time scale, namely, $R_c < 0.1$ K/min. Conversely, θ-TlCo exhibits the CO transition even upon cooling at a rate of 150 K/min. However, the symptom of the CG observed in the X-ray diffuse scatterings lead us to assume that the CG would come out when θ-TlCo can be cooled faster; namely, $R_c > 150$ K/min, which is several orders of magnitude larger than that of θ-RbZn.

Now, we can see the relationship between charge frustration and charge-glass-forming ability. The physical properties of θ-CsZn, θ-RbZn and θ-TlCo are listed in Table 3.1. There is a clear tendency that the larger charge frustration leads to the lower R_c value, i.e., higher charge-glass-forming ability. Charge frustration

Fig. 3.24 Schematic of the cooling-rate-dependent bifurcation of the charge liquid at high temperatures into the charge order and the charge glass at low temperatures

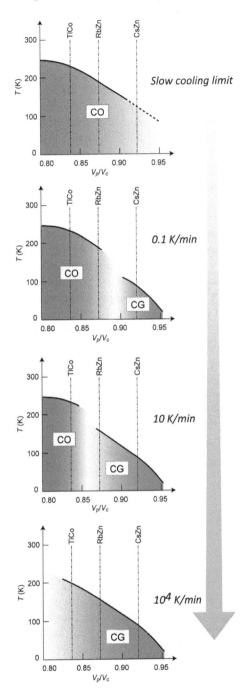

generally works against the formation of a particular charge configuration; in other words, it impedes the growth of the charge ordering, thus reducing the value of R_c. Notably, R_c is extremely sensitive to the degree of charge frustration. It is noted that, while the CO transition temperature does not considerably differ between of θ-RbZn (195 K) and θ TlCo (245 K), the R_c values differ by more than one order of magnitude. Finally, we mention the effect of electron-lattice (e-l) coupling. Considering the lattice modulation accompanied by the first-order CO transition and the observation of X-ray diffuse scattering characterized by particular CO wave vectors, the e-l interactions are expected to play some role in the CO and CG formation, possibly the slowing down of the charge fluctuations. Although it is difficult to examine whether the CG states appear in the absence of the e-l coupling in real systems, the relevance of the degree of the electronic frustration to the glass-forming ability suggests that the primary origin of the CG formation is electronic. We also note a recent theoretical simulation based on the frustrated triangular lattices, which reproduces a CG behavior even without the e-l coupling and disorder, indicating the importance of charge frustration [36].

As seen above, the charge frustration is a controlling parameter that dominates the CG characteristics. Furthermore, our results also verify that the cooling rate is also a significant control parameter in determining the low-temperature electronic state. These findings enable us to construct a comprehensive diagram, including the equilibrium and nonequilibrium phases, with the three parameters of temperature, charge frustration, and the cooling speed, as shown in Fig. 3.24. The notion that geometrical frustration is of significance for determining the kinetic aspects of a phase transition (such as R_c) as well as the static aspects (such as transition temperature) can be applied to much broader classes of electronic systems with competing phases.

3.5   Summary

In this chapter, we have investigated the charge-order and/or charge-glass behaviors of the three members in the θ-(BEDT-TTF)$_2$X family, θ-CsZn, θ-RbZn and θ-TlCo, with different charge frustration. First, the measurements of resistivity, resistance noise, and X-ray diffuse scatterings for the most frustrated system, θ-CsZn, found several hallmarks of glass in charge sector, demonstrating the occurrence of charge-glass state. The material dependence of the glassy behavior has also provided insights into the nature of the charge-glass state; charge frustration is a dominating factor for the glassy freezing and may even affect the collective nature of the charge dynamics. It is a prime conclusion in this chapter that the charge glass shares common fundamentals in physics with the classical structural glass. Note that inhomogeneous electronic states are often found in strongly correlated electron systems, such as manganites [37, 38] and high-transition-temperature cuprates [39, 40]. However, they are inevitably affected by disorders, which is argued to play a key role in the emergence of the inhomogeneity. Unlike that, the organic systems investigated here are basically disorder-free and thus the present charge-glass phenomena can be distinguished from the disorder-induced phenomena.

The present study has an additional consequence that stronger charge frustration leads to smaller critical cooling rate, R_c, that is, superior charge-glass former. In other words, charge frustration slows down the kinetics that drives a transition to the charge ordering.

References

1. P. Debenedetti, F. Stillinger, Nature **410**, 259 (2001)
2. F. Nad, P. Monceau, H.M. Yamamoto, J. Phys. Condens. Matter **20**, 485211 (2008)
3. R. Brüning, K. Samwer, Phys. Rev. B **46**, 11318 (1992)
4. I.M. Hodge, Science **267**, 1945 (1995)
5. M.D. Ediger, C.A. Angell, S.R. Nagel, J. Phys. Chem. **100**, 13200 (1996)
6. C.A. Angell, K.L. Ngai, G.B. McKenna, P.F. McMillan, S.W. Martin, J. Appl. Phys. **88**, 3113 (2000)
7. J. Müller, Chem. Phys. Chem. **12**, 1222 (2011)
8. C. Gainaru, S. Kastner, F. Mayr, P. Lunkenheimer, S. Schildmann, H.J. Weber, W. Hiller, A. Loidl, R. Böhmer, Phys. Rev. Lett. **107**, 118304 (2011)
9. P. Dutta, P. Dimon, P.M. Horn, Phys. Rev. Lett. **43**, 646 (1979)
10. P. Dutta, P.M. Horn, Rev. Mod. Phys. **53**, 497 (1981)
11. M.B. Weissman, Rev. Mod. Phys. **60**, 537 (1988)
12. M.B. Weissman, R.D. Black, P.J. Restle, T. Ray, Phys. Rev. B **27**, 1428 (1983)
13. S. Scouten, Y. Xu, B.H. Moeckly, R.A. Buhrman, Phys. Rev. B **50**, 16121 (1994)
14. J. Müller, J. Brandenburg, J.A. Schlueter, Phys. Rev. B **79**, 214521 (2009)
15. S. Machlup, J. Appl. Phys. **25**, 341 (1954)
16. T. Mori, Bull. Chem. Soc. Jpn. **73**, 2243 (2000)
17. H. Tajima, S. Kyoden, H. Mori, S. Tanaka, Phys. Rev. B **62**, 9378 (2000)
18. A. Widmer-cooper, P. Harrowell, H. Fynewever, Phys. Rev. Lett. **93**, 135701 (2004)
19. J.P.K. Doye, D.J. Wales, F.H.M. Zetterling, M. Dzugutov, J. Chem. Phys. **118**, 2792 (2003)
20. H. Tanaka, T. Kawasaki, H. Shintani, K. Watanabe, Nat. Mater. **9**, 324 (2010)
21. H. Shintani, H. Tanaka, Nat. Phys. **2**, 200 (2006)
22. Y. Nogami, J.-P. Pouget, M. Watanabe, K. Oshima, H. Mori, S. Tanaka, T. Mori, Synth. Met. **103**, 1911 (1999)
23. M. Watanabe, Y. Nogami, K. Oshima, H. Mori, S. Tanaka, J. Phys. Soc. Jpn. **68**, 2654 (1999)
24. R. Chiba, K. Hiraki, T. Takahashi, H.M. Yamamoto, T. Nakamura, Phys. Rev. Lett. **93**, 19 (2004)
25. R. Chiba, K. Hiraki, T. Takahashi, H.M. Yamamoto, T. Nakamura, Phys. Rev. B **77**, 1 (2008)
26. C.A. Angell, Science **267**, 1924 (1995)
27. R. Böhmer, K.L. Ngai, C.A. Angell, D.J. Plazek, J. Chem. Phys. **99**, 4201 (1993)
28. M. Watanabe, Y. Noda, Y. Nogami, H. Mori, J. Phys. Soc. Jpn. **73**, 116 (2004)
29. K. Suzuki, K. Yamamoto, K. Yakushi, Phys. Rev. B **69**, 85114 (2004)
30. H. Mori, S. Tanaka, T. Mori, Phys. Rev. B **57**, 12023 (1998)
31. M. Naka, H. Seo, J. Phys. Soc. Jpn. **83**, 53706 (2014)
32. P.T. Sarjeant, R. Roy, Mater. Res. Bull. **3**, 265 (1968)
33. H. Tanaka, J. Non. Cryst. Solids **351**, 3371 (2005)
34. A. Takeuchi and A. Inoue, Mater. Sci. Eng. **304**, 446 (2001)
35. C. Tang, P. Harrowell, Nat. Mater. **12**, 507 (2013)

36. S. Mahmoudian, L. Rademaker, A. Ralko, S. Fratini, V. Dobrosavljević, Phys. Rev. Lett. **115**, 1 (2015)
37. E. Dagotto, Science **309**, 257 (2005)
38. Y. Tokura, Rep. Prog. Phys. **69**, 797 (2006)
39. I. Zeljkovic, Z. Xu, J. Wen, G. Gu, R.S. Markiewicz, J.E. Hoffman, Science (80-.). **337**, 320 (2012)
40. I. Raičević, J. Jaroszyński, D. Popović, C. Panagopoulos, T. Sasagawa, Phys. Rev. Lett. **101**, 177004 (2008)

Chapter 4
Electronic Crystal Growth

Abstract In this chapter, we focus not on the charge-glass state itself but on the transforming process from charge glass into charge order, that is, electronic crystal growth. First, time dependence of resistivity during the process is measured. We found that obtained crystallization time can be well described by dome-like structure called Time-Temperature-Transformation (T-T-T) curve. It provides a strong evidence for the existence of two distinct processes in electronic crystal growth; nucleation and growth. Furthermore, NMR measurements, which serve as a probe of local charge density profile, give us the remarkable data that the intermediate states different from both initial charge-glass and final charge-ordered states are formed before the formation of true charge ordering near T_g. It is reminiscent of two-step nucleation discussed intensively in classical colloidal system, indicating the universality of such novel nucleation process.

Keywords θ-(BEDT-TTF)$_2$X · Charge order · Charge glass · Crystallization · Crystal nucleation · Crystal growth · Time-temperature-transformation curve · Two-step nucleation

4.1 Introduction

In Chap. 3, θ-(BEDT-TTF)$_2$X family has been established to be a typical CG former. We then focus on a transient process from the CG state into a CO state, that is, electronic crystallization.

Needless to say, a crystal is the most fundamental form of order in interacting many-body systems. Atoms, molecules, polymers, colloids and so on form crystals via their mutual interactions. The process of crystallization in classical systems has been a seminal issue of broad interest, ranging from basic science to applications such as material processing. It is one of the most major research fields in the science of condensed matter [1, 2]. Experimentally, the crystallization of atoms or

© Springer Nature Singapore Pte. Ltd. 2017   63
T. Sato, *Transport and NMR Studies of Charge Glass in Organic
Conductors with Quasi-triangular Lattices*, Springer Theses,
DOI 10.1007/978-981-10-5879-0_4

molecules is observable as a time-dependent phenomenon that occurs from a supercooled liquid or glass [3]. In a system of electrons, the electrons interact repulsively and can form an electronic crystal called the Wigner crystal [4]. Even in a solid, an electronic crystal with a periodicity distinct from that of the underlying lattice has been widely observed in the form of CO in strongly correlated electronic systems [5]. Electronic crystallization can be considered as a process in which electrons lose wave nature and acquire particle nature. Does the crystallization in the quantum system proceed in the same manner as in conventional classical systems? Here, we report the first successful observation of the process of electronic crystallization and provide an answer to this open question.

It is the layered organic material θ-RbZn that makes it possible to experimentally access this issue. Distinctive from θ-CsZn that shows CG but no CO even when slowly cooled, the resultant electronic states in θ-RbZn can be easily tuned by cooling speed applicable in laboratory (Figs. 1.6 and 4.1); it exhibits an electronic crystal (T_{CO} = 198 K) only when slowly cooled (<1 K/min) [6–8], and when cooled rapidly (>5 K/min), the transition is bypassed to produce a supercooled state [9], followed by a CG transition at approximately T_g = 160 K. The appropriate value of frustration in θ-RbZn sets its crystallization rate in the laboratory time scale. Therefore, θ-RbZn serves as a suitable platform for investigating electronic crystallization on an experimentally accessible time scale.

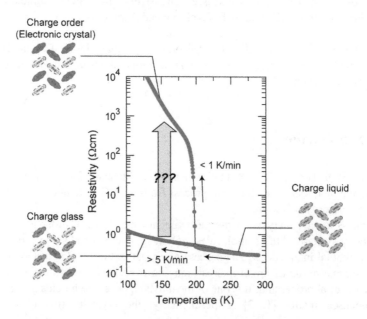

Fig. 4.1 The temperature dependence of resistivity for θ-RbZn at different cooling rates

4.2 Experimental

Notably, an electronic crystal and glass (or supercooled liquid) can be probed by charge- and spin-sensitive measurements, distinctively from the case of the atomic or molecular systems. The resistivity is much lower in the supercooled electronic liquid/glass state than in the electronic crystal state and thus characterizes the electronic state macroscopically. Moreover, to examine the crystallization process in molecular scale, we exploit ^{13}C nuclear magnetic resonance (NMR), which is capable of distinguishing the electronic liquid/glass and crystal states through detection of local charge/spin disproportionation. In the present work, θ-RbZn was quenched (4–6 K/min) to the target temperature, T_q, from above T_{CO} to prepare a supercooled or glass state. Next, the time evolution of the electronic state during electronic crystallization was investigated by monitoring the resistivity and NMR spectrum while keeping the temperature at T_q to unveil the process of formation of the electronic crystal.

4.3 Time-Temperature-Transformation Curve in Resistivity Growth

First, we performed resistivity measurements during the electronic crystal growth, and reveal the overall characteristics in it. Figure 4.2a displays the time dependence of resistivity at several temperatures. Each data in Fig. 4.2a are normalized by the initial value of resistivity, $\rho(t = 0)$, and the saturated value, $\rho(t = \infty)$, which is determined by the resistivity in a CO state reached when slowly cooled. At $T_q = 191$ K (just below T_{CO}), it takes more than 10^4 s before the crystallization starts; however, once it begins, it is completed rapidly. When T_q is lowered to 188 K and subsequently to 180 K, the onset time becomes shortened. When T_q is lowered further, however, the onset time becomes longer, and the profile of the time evolution of the resistivity changes such that the crystallization proceeds gradually after it sets in. From Fig. 4.2a, we define the crystallization time (t_{cry}) as the time at which resistivity reaches a fixed fraction of the saturated values. As the criterion, we use a particular percentages of 1, 5, 10, 20, 30, 40, and 50% of the saturated resistivity value. The plot of t_{cry} versus T_q for several fixed fractions is displayed in Fig. 4.2b, and we found the dome-like structures. The curve of such a shape has been widely observed for the crystallization of structural glasses in metallic alloys [10] and is called the time-temperature-transformation (TTT) curve. The present results provide the first evidence for the universality of the TTT curve for electronic crystallization.

Here, we briefly review the classical theory for the TTT curve. The TTT curve for the structural crystallization is satisfactorily described by classical nucleation theory, which is based on the two processes governing crystallization; nucleation,

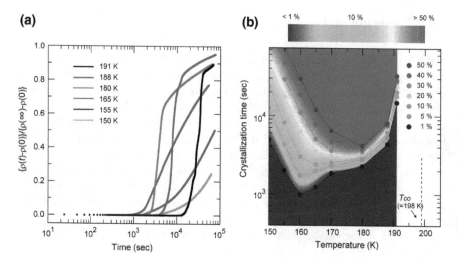

Fig. 4.2 Time evolution of electronic crystallization from a supercooled liquid or a glass and the temperature profile of crystal growth. **a** The time dependence of resistivity measured at different T_q after rapid cooling across T_{CO}. The plot for each temperature is normalized by the initial value, $\rho(t = 0)$, and the saturated value, $\rho(t = \infty)$, which is determined by the resistivity in a CO state under slow cooling. **b** The time-temperature-transformation (TTT) diagram of θ-RbZn determined from (**a**). The range of *color* represents the percentage of resistivity recovery. The *circles* indicate the t_{cry} for particular percentages (1, 10, 20, 30, 40, and 50%) in resistance recovery

and growth [11, 12]. In the first step, small nuclei are generated in a supercooled liquid or a glass through structural fluctuations at a rate given by

$$I = I_0 k_v \exp\left[-\frac{\Delta G^*}{k_B T}\right], \qquad (4.1)$$

where I_0 is a constant, k_v is the kinetic factor, and ΔG^* is the free-energy barrier for the formation of a critical nucleus [13]. The key parameters to determine ΔG^* are the crystal/liquid interface energy cost, γ, against nucleation and the free-energy difference between crystal and liquid, $\Delta\mu$, which promotes nucleation as a thermodynamic driving force and increases as T is lowered. ΔG^* is known to be expressed in the form of $\frac{16\pi\gamma^3}{3\Delta\mu^2}$ [14]. The kinetic factor k_v, which measures the diffusivity in the configurational space in the glass state, decreases in magnitude toward low temperature. In the second process, the formed nuclei increase in size at a rate given by

$$V = V_0 k_v \left(1 - \exp\left[-\frac{v_p \Delta\mu}{k_B T}\right]\right), \qquad (4.2)$$

where V_0 is a constant, and v_p is the volume per particle [15]. The general temperature profiles of I and V expected from these equations are shown in Fig. 4.3,

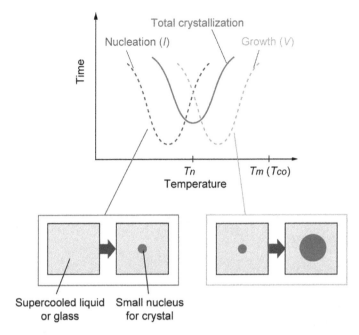

Fig. 4.3 Schematics of the crystallization rate as a function of temperature. The speed of crystallization (*red curve*) is determined by the combination of nucleation (*blue dashed-curve*) and growth (*yellow dashed-curve*) processes. The images if nucleation and growth processes are drown in the insets

such that the growth rate reaches a maximum at a higher temperature than that of the nucleation rate [16, 17]. As a result, the total crystallization rate reaches a maximum in between the two domes (Fig. 4.3). Therefore, observed TTT curve during electronic crystal growth can be interpreted as a strong evidence for the existence of the corresponding nucleation and growth process in it. It is noted, moreover, that the maximum crystallization rate at the "nose temperature, T_n", signifies a crossover between nucleation- and growth-dominated regimes. The T_n for the present case corresponds to approximately 170–180 K in the TTT curves for the tens of percent resistance recovery that is indicative of sizable crystallization (Fig. 4.2b). This explains the distinctive time evolutions of resistivity above and below the T_n (Fig. 4.2a); for $T > T_n$, the abrupt resistivity increase after a long quiescent time indicates that the nucleation (as a rare event) is immediately followed by a rapid growth of the crystal seeds, whereas the gradual resistivity increase for $T < T_n$ suggests that the growth rate (which is suppressed under the nucleation rate at low T) regulates the speed of crystallization. The two regimes of crystal growth above and below T_n are also clearly distinguished in the change rate of color in the contour plot (Fig. 4.2b). Noticeably, the T_n nominally shifts from 170 to 180 K in the TTT curve of the 50% resistance recovery to 160 K for that of the 1% resistance recovery. The initial resistivity increase is seemingly caused by nucleation. Given that it is measured by the 1% TTT curve, the nucleation rate is suggested to peak at approximately 160 K, whose value of T is lower than T_n for the total crystallization, in accordance with the general behavior of structural glasses (Fig. 4.3).

These results also imply the followings; in the former regime, macroscopic inhomogeneity likely occurs owing to the occurrence of well-defined CO domains; indeed, we often encountered irregular time evolutions of resistivity suggestive of the macroscopic inhomogeneity above T_n. Below T_n, in contrast, the time evolution of resistivity was always smooth as expected for finely distributed microcrystals, which are expected to behave like a homogeneous effective medium.

4.4 Formation of Intermediate State Before CO Nucleation

Next, we examine the crystallization process through investigation of the ^{13}C-NMR spectra, which serves as a probe of the charge density profile. The magnetic field of 6 T was applied in a particular direction to form the so-called "magic angle" with the ^{13}C=^{13}C vector in the ^{13}C-enriched BEDT-TTF molecule and was set to be parallel to the ab plane to avoid the unwanted complications of the NMR spectra arising from nuclear dipolar splitting and molecular inequivalence against the field direction (see Chap. 2). At room temperature, the spectrum has a two-line structure (Fig. 4.4a) originating from the two adjacent ^{13}C sites in the center of the molecule. As T is lowered, the lines become broadened because of the slowing down of charge fluctuations (Fig. 4.4b). When T slowly passes through T_{CO}, the spectrum exhibits the structural characteristics of a CO state (Fig. 4.4c); the complicated structure is decomposed into two components arising from the charge-rich and - poor molecules [18] (see Chap. 2). When the system is rapidly cooled to T_q through T_{CO}, however, the spectrum exhibits a broad feature indicative of a CG state with spatially inhomogeneous charge density. Next, the spectrum was traced as a function of time at each T_q. The case of $T_q = 140$ K is shown in Fig. 4.4d; the broad spectrum observed just after cooled to 140 K gradually changes its structure with time and eventually takes the shape characteristic of the CO state. This is the first microscopic observation of the electronic crystal growth from a glass state.

We then deduce the time dependence of the volume fraction of the CO domains by decomposing the spectrum at each time into the CG and CO components with the relative intensity as a fitting parameter. We explain the fitting procedure by taking the data for $T_q = 140$ K as an example. Figure 4.5a shows the spectrum observed immediately after the sample is rapidly (at a rate of 4 K/min) cooled to 140 K (Solid curve), which can be regarded as the spectra of the charge glass state without CO fraction, whereas the saturated spectrum observed at sufficiently long times after the cooling (Dashed curve) represents the spectrum of a 100% CO fraction. The former and latter spectral profiles are denoted by $I_{0\%}$ and $I_{100\%}$,

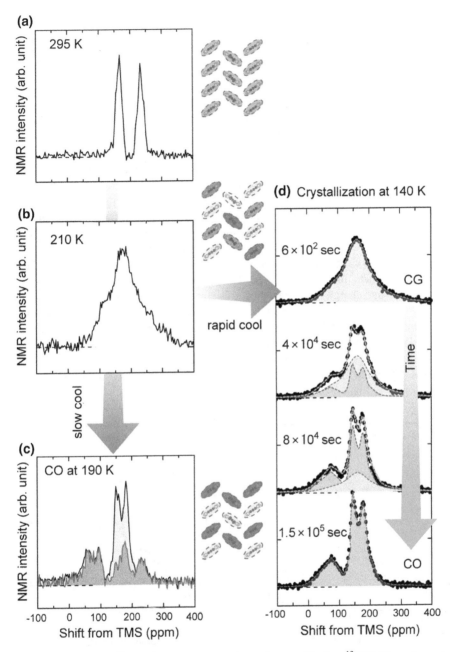

Fig. 4.4 Evolution of ^{13}C NMR spectra during electronic crystallization. ^{13}C-NMR spectra **a** at room temperature and **b** at 210 K. **c** ^{13}C-NMR spectra of a CO state obtained at 190 K after slowly cooled from above T_{CO}. **d** The time evolution of ^{13}C-NMR spectra during the crystallization process at 140 K. The *blue* and *red* parts indicate the CG and CO components fitting the spectra, respectively. The sum of the two components in the fitting (*broken lines*) reproduces the observed spectra

respectively, and each spectrum observed at a time of t, $I(t)$, during crystallization is fitted by the form of

$$I(t) = (1 - A) \cdot I_{0\%} + A \cdot I_{100\%} \qquad (4.3)$$

where the fitting parameter, A, gives the volume fraction of CO domains at t. A typical result is shown in Fig. 4.5b (at $t = 8 \times 10^4$ s for $T_q = 140$ K), in which the red dashed line shows a good fit to the data obtained with $A = 0.7$. Figure 4.6 shows the deduced volume fraction of CO domains versus time during crystallization.

Now, we discuss the time evolution of the CO-domain fraction at 191 K (in the nucleation-dominated regime) and at 150 K (in the growth-dominated regime) (Figs. 4.7a, c). Remarkably, the two curves are distinctive in relation to the time evolution of conductivity. At 191 K, the CO-domain fraction rapidly increases in parallel with a steep decrease in conductivity, consistent with the macroscopically inhomogeneous growth of the CO domains (as depicted in Fig. 4.7b) as discussed above. At 150 K, however, the CO domains remain undeveloped while the conductivity shows a significant decrease during the period of 400–2000 s (highlighted by yellow shade in Fig. 4.7c). The ^{13}C-NMR can probe locally formation of charge-rich and charge-poor sites, thus, this puzzling behavior suggests that while a considerable fraction of the sample becomes highly resistive, this fraction is not the CO domains. A clue for resolving this puzzle is found in the recent study of colloidal hard-sphere systems, which revealed that metastable intermediate domains different from both of glass and crystalline phases are formed before the formation of true crystallite [19, 20] (Fig. 4.7d). The alternative pathway distinctive from a

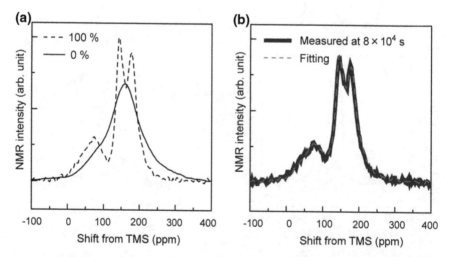

Fig. 4.5 Typical fitting results of ^{13}C-NMR spectra during crystallization. **a** NMR spectra in the glass state observed immediately ($t = 2 \times 10^3$ s) after the quenching to $T_q = 140$ K (*solid curve*) and in the CO state observed at a sufficiently long time $t = 1 \times 10^5$ s from the quenching (*dashed curve*). **b** The *red dashed line* is a fit to the data using Eq. (4.3)

Fig. 4.6 Time evolution of
the CO domains at 140 K.
Time dependence of the
parameter A in the fitting form
of the spectra [Eq. (4.3)]
during crystallization at
140 K

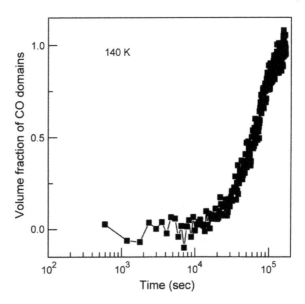

conventional one-step nucleation, that is, the direct formation of the thermody-
namically stable crystalline nuclei, is often referred as "two-step nucleation".
Although the intermediate state, whose structure is between true crystal and glass,
cannot be a thermodynamic ground state, it likely can work as a buffer and reduce
the interface energy cost γ, decreasing the free-energy barrier for nucleation,
$\Delta G^* \left(= \frac{16\pi\gamma^3}{3\Delta\mu^2} \right)$. The schematic nucleation free-energy profiles satisfactorily describe
the one- and two-step pathway for crystal phase (Fig. 4.8). The two-step pathway is
likely preferable owing to the lower free-energy cost. In fact, this is an example of
the well-known Ostwald's step rule, which states that transitions from one phase to
another occur through the lowest free-energy pathway. The same mechanism can be
applied for the electronic crystallization; below T_g, where charge fluctuations are
nearly frozen in the present case, the formation of CO domains, which is formed at
the highly cost of the interface energy, is less promoted than at higher temperatures.
As a result, the system seeks an alternative pathway as a means to complete to
crystalize, that is, the formation of intermediate state. In the present system, the
metastable intermediate domains are highly resistive but show no clear difference in
the NMR spectra from the glass state. Because the NMR spectra reflect the charge
disproportionation [6], the results indicate that the disproportionate charge density
is not appreciably developed in the intermediate state. Instead, the spatial configuration of charge density is probably changed in this state. Subsequently, the CO
domains emerge after the intermediate domains grow considerably (Fig. 4.7c, d).
The present results suggest the novel two-step nucleation mechanism occurs at low
temperatures below T_n.

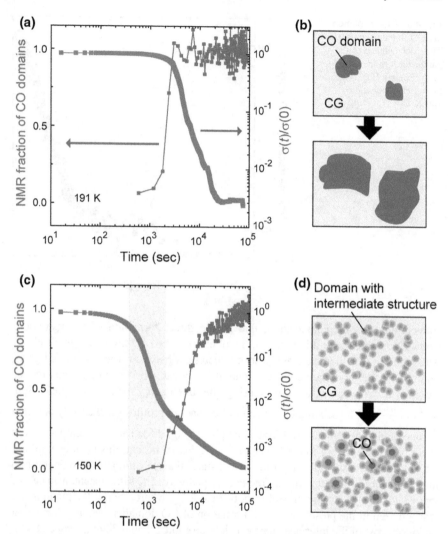

Fig. 4.7 Comparison of the time evolution of conductivity and NMR spectra toward electronic crystallization. **a** and **c** Time dependences of the CO-domain fraction (*red*), which is deduced from the NMR spectra, and conductivity (*blue*) at (**a**) 191 K (the nucleation-dominated regime) and (**c**) 150 K (the growth-dominated regime). **b** and **d** Schematics of the CO-domain evolution in (**b**) the high-T nucleation-dominated regime, and (**d**) the low-T growth-dominated regime

Fig. 4.8 The schematic figure of one- and two-step pathway towards crystal phase (obtained from Ref. [21])

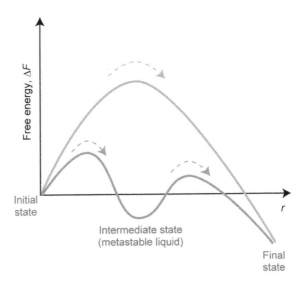

4.5 Summary

In this Chapter, we report the first observation of the crystal growth of electron in charge-glass hosting system, θ-RbZn, and revealed the mechanism of this electronic crystal growth by the combination of resistivity and NMR measurements, which are unavailable for conventional structural glasses. They provide not only the inherent characteristics of crystallization in electrons but also the additional insight for generic glass physics.

The process of electronic crystallization revealed here is summarized as follows. Overall, it consists of two distinct processes; nucleation and growth. In line with the classical crystallization theory, the metastable supercooled electronic liquid that appears at high temperatures above 180 K becomes crystallized by overcoming the free-energy barrier against the crystal. However, electronic crystallization from the non-equilibrium glass state at lower temperatures occurs in a different manner; because it cannot overcome the free-energy barrier with insufficient thermal energy, it seeks an alternative pathway to crystallization via the formation of the intermediate phase. The present study demonstrated that the crystal growth process is no longer limited to the formation of condensed matter of atoms and molecules but is now extended to electronic matter. Thus, we have a novel interdisciplinary platform that integrates the science of correlated electrons and the science of soft matter.

References

1. T. Palberg, J. Phys. Condens. Matter **11**, R323 (1999)
2. U. Gasser, J. Phys. Condens. Matter **21**, 203101 (2009)
3. U. Gasser, E.R. Weeks, A. Schofield, P.N. Pussey, D.A. Weitz, Science **292**, 258 (2001)
4. E. Wigner, Phys. Rev. **46**, 1002 (1934)
5. E. Dagotto, Science **309**, 257 (2005)
6. K. Miyagawa, A. Kawamoto, K. Kanoda, Phys. Rev. B **62**, R7679 (2000)
7. K. Yamamoto, K. Yakushi, K. Miyagawa, K. Kanoda, A. Kawamoto, Phys. Rev. B **65**, 85110 (2002)
8. F. Nad, P. Monceau, H.M. Yamamoto, Phys. Rev. B **76**, 205101 (2007)
9. M. Watanabe, Y. Noda, Y. Nogami, H. Mori, Synth. Met. **135–136**, 665 (2003)
10. J.F. Löffler, J. Schroers, W.L. Johnson, Appl. Phys. Lett. **77**, 681 (2000)
11. P.G. Debenedetti, *Metastable Liquids: Concepts and Principles* (Princeton University Press, 1996)
12. D. Turnbull, Contemp. Phys. **10**, 473 (1969)
13. D. Turnbull, J.C. Fisher, J. Chem. Phys. **17**, 71 (1949)
14. K.F. Kelton, Solid State Phys. **45**, 75 (1991)
15. D.R. Uhlmann (Plenum, New York, 1969)
16. H. Senapati, K.K. Kadiyala, C.A. Angell, J. Phys. Chem. **95**, 7050 (1991)
17. J. Orava, A.L. Greer, B. Gholipour, D.W. Hewak, C.E. Smith, Nat. Mater. **11**, 279 (2012)
18. R. Chiba, K. Hiraki, T. Takahashi, H.M. Yamamoto, T. Nakamura, Phys. Rev. Lett. **93**, 19 (2004)
19. S. Iacopini, T. Palberg, H.J. Schöpe, J. Chem. Phys. **130** (2009)
20. J.R. Savage, A.D. Dinsmore, Phys. Rev. Lett. **102**, 198302 (2009)
21. E. Sanz, C. Valeriani, Nat. Mater. **14**, 15 (2014)

Chapter 5
Conclusions

Abstract This thesis demonstrates the concept of the new class of glass in electrons in organic conductors. In Chap. 3, typical hallmarks of glass are clearly confirmed, giving strong evidences for the presence of charge glass. Material dependence also provides an insight for the origin of charge-glass state; geometrical frustration not disorder plays a significant role in forming charge glass, which is distinctive from previous inhomogeneous electronic states in strongly correlated electron systems. In Chap. 4, we reported T-T-T curve during electronic crystal growth, strongly indicating the existence of nucleation and growth process in line with classical crystal growth. Moreover, near T_g, the novel two-step nucleation is revealed by NMR measurements. This is the first macro- and micro-scopic observation of electronic crystal growth. All these results provide the important statement that glassy behavior is no longer limited in classical soft materials but is extended quantum hard materials. Series of my study uncovers organic conductor θ-(BEDT-TTF)$_2$X serves as a novel interdisciplinary platform uniting the strongly correlated electron system and soft matter, and opens up new possibilities in the field of glassiness.

Keywords Strongly correlated electrons · θ-(BEDT-TTF)$_2$X · Geometrical frustration · Charge order · Charge glass · Elecronic crystal growth · Time-Temperature-Transformation curve · Two-step nucleation

In this thesis, we explored the quasi-two-dimensional organic conductor θ-(BEDT-TTF)$_2$X with anisotropic triangular lattices. The nature of charge ordering in this system is strongly coupled with geometrical frustration originating from the triangular lattices, potentially giving rise to unconventional electron states without long-ranged order. Motivated by this scenario, we investigated the θ-(BEDT-TTF)$_2$X family of materials in the light of glassiness in electrons.

We first focused on θ-CsZn with the nearly isotropic triangular lattice, that is, the highly frustrated system, and confirmed the occurrence of a novel glass state in electrons by demonstrating several hallmarks of glass in charge sector: (i) non-equilibrium, (ii) slow dynamics, and (iii) short-ranged correlation. The

© Springer Nature Singapore Pte. Ltd. 2017
T. Sato, *Transport and NMR Studies of Charge Glass in Organic Conductors with Quasi-triangular Lattices*, Springer Theses,
DOI 10.1007/978-981-10-5879-0_5

cooling-rate dependence and physical aging in resistivity clearly shows the emergence of non-equilibrium charge dynamics. In addition, noise measurements revealed that the temperature dependence of the relaxation time related to charge fluctuations follows the Arrhenius-type function; thus, the present glass-forming system can be categorized as a "strong glass", indicating that the fluctuations are associated with the local rearrangement of charge configurations. The phenomenological analysis based on the DDH model also supports the picture of the strong glass, resolving the precise structure of activation energies peaking at a particular value of energy scale, which is consistent with the observed Arrhenius-type relaxation. Moreover, X-ray diffuse scattering measurements revealed that the spatial growth of the "3 × 3"-period CO domains is frozen at the glass transition point. All these experimental observations showed that θ-CsZn is unequivocally a CG former.

Next, we investigated the analogous materials with the different strengths of frustration; θ-RbZn and θ-TlCo, whose crystal symmetry is equal to θ-CsZn, but the anisotropy of their triangular lattices is varied. Consequently, the order of the frustration strength is θ-CsZn (highest frustrated) > θ-RbZn > θ-TlCo (least frustrated). By comparing the properties of charge-glass states in θ-RbZn with those in θ-CsZn, we found the "fragility" of θ-RbZn to be slightly higher than that of θ-CsZn, suggesting that the charge frustration is coupled with the collective nature of electrons. This behavior is consistent with the theoretical suggestions in a classical glass, thus implying the universal correlation between geometrical frustration and nature of glass. By investigating the least frustrated system, θ-TlCo, we demonstrated that the systematic variations in the charge-glass-forming ability among θ-TlCo, θ-RbZn, and θ-CsZn are in good agreement with those in charge frustration. Therefore, we conclude that charge frustration tends to slow the charge-ordering kinetics and leads to a superior charge-glass former.

Finally, we investigated the transformation process from a charge glass into a charge order, namely, electronic crystal growth, by means of resistivity and NMR measurements. We observed TTT curve, strongly indicating the existence of nucleation and growth processes during electronic crystal growth, in line with classical crystal growth. Moreover, we compared the time dependence of the NMR spectra and the resistivity, corroborating the so-called two-step nucleation mechanism near T_g. Our results are the first macro- and microscopic observations of the crystallization process in electrons. All these results provide an important insight that glassy behavior is no longer limited to classical soft materials but is extended to quantum hard materials.

Our series of studies uncovered that the organic conductor θ-(BEDT-TTF)$_2$X serves as a novel interdisciplinary platform that unites a strongly correlated electron system with soft matter, which opens up new possibilities in the field of glassiness.

About the Author

Takuro Sato studied physics in condensed materials and received his Ph.D. from the Department of Applied Physics, The University of Tokyo in 2017. His research interest is the physical properties of organic conductors of θ-(BEDT-TTF)$_2$X family, which involves both a charge-ordering instability and a frustration in their lattice geometry. He investigated the series of materials by various measurements: resistivity, time-resolved electric transport, X-ray diffraction analysis, and nuclear magnetic resonance spectroscopy, and successfully observed for the first time the apparent hallmarks of a so-called charge glass, which has been intensively debated theoretically and experimentally. In April 2017, he began to work as a Post-Doc in the group of Fumitaka Kagawa at RIKEN Center for Emergent Matter Science.

© Springer Nature Singapore Pte. Ltd. 2017
T. Sato, *Transport and NMR Studies of Charge Glass in Organic
Conductors with Quasi-triangular Lattices*, Springer Theses,
DOI 10.1007/978-981-10-5879-0

CPSIA information can be obtained
at www.ICGtesting.com
Printed in the USA
LVHW02*1748180318
570251LV00001B/3/P